FLEXIBILITY IN ADAPTATION PLANNING: WHEN TO INCLUDE FLEXIBILITY FOR INCREASING RESILIENCE

FLEXIBILITY IN ADAPTATION PLANNING: WHEN, WHERE AND HOW TO INCLUDE FLEXIBILITY FOR INCREASING URBAN FLOOD RESILIENCE

DISSERTATION

Submitted in fulfilment of the requirements of

the Board for Doctorates of Delft University of Technology

and

of the Academic Board of the UNESCO-IHE

Institute for Water Education

for

the Degree of DOCTOR

to be defended in public on

Wednesday, 22nd November 2017, at 10:00 hours

in Delft, the Netherlands

by

Mohanasundar RADHAKRISHNAN

Master of Science in Water Supply Engineering UNESCO-IHE Institute for Water Education

born in Palayamkottai, India

This dissertation has been approved by the

promotor: Prof. dr. C. Zevenbergen

copromotor: Dr. A. Pathirana

Composition of the doctoral committee:

Chairman	Rector Magnificus TU Delft
Vice-Chairman	Rector UNESCO- IHE
Prof. dr. C. Zevenbergen	UNESCO- IHE / TU Delft, promotor
Dr. A Pathirana	IHE Delft, copromotor

Independent members:

Prof. dr. T.J.M. Spit	Universiteit Utrecht
Prof. dr. ir. M. Kok	TU Delft
Prof. dr. R. Ranashinge	UNESCO- IHE / TU Twente
Prof. dr. Srikantha Herath,	United Nations University, Tokyo
Prof. dr. ir. W.A.H. Thissen	TU Delft, reserve member

Prof. Richard Ashley and Dr. Berry Gersonius from UNESCO- IHE, as the Project leader and Project Manager respectively of CRCWSC Project B 4.2, have contributed significantly to the development of this dissertation. This research was funded by Cooperative Research Centre for Water Sensitive Cities (CRCWSC), which is a part of the Australian Commonwealth Government's Cooperative research centre (CRC) programme.

This research was conducted under the auspices of the Graduate School for Socio-Economic and Natura Sciences of the Environment (SENSE)

Published by:

CRC Press/Balkema

Schipholweg 107C, 2316XC, Leiden, The Netherlands

Pub.NL@taylorandfrancis.com

www.crcpress.com – www.taylorandfrancis.com

ISBN 978-0-8153-5729-2

This thesis is dedicated to my guardian angels in heaven

Gunapakkiam, my grandmother

And

Raman sir, my Engineering guru

Acknowledgements

எந்நன்றி கொன்றார்க்கும் உய்வுண்டாம் உய்வில்லை
செய்ந்நன்றி கொன்ற மகற்கு

Verse 110 Tirukural (Thiruvalluar 31 BC)

Forgetting to do good and you may find salvation yet

But forget gratitude you're headed for destruction's net

Translation by Gandhi (2015)

I am overwhelmed with the support and wishes of so many nice and kind people throughout my life. At the outset I thank the people and the Commonwealth Government of Australia for supporting my research through the Cooperative Research Centre for Water Sensitive Cities project (CRCWSC). I thank Dr. S.D. Badri Nath and Dr. A. Panner Selvam for their guidance and support in encouraging me to pursue research. I thank my friend Kathiravan Jeyakumar for the Tamil translation and Mr. Geor Hintzen for the Dutch translation of the summary. I also remember with thank the continuous support of Mr. Naga Sreenivas and Dr.Mohanavadivu for their inputs since the proposal stage of this research. I thank Prof. Arthur Mynett – the Saturday morning chats - and Prof. Rosh Ranasinghe for their words of encouragement throughout the thesis period. I thank Jolanda, Anique, Tonneke and Martine for the wonderful administrative support through my stay at IHE. My ultimate Alpha team of "*Mike Ones*" at GIZ – especially, Esakki, Ramesh Ji, Jiten, Vibhor, Venu – shared their invaluable day to day knowledge and difficulties on urban adaptation.

Dr. Tushith Islam contributed not only to the development of context specific adaptation grammar but also introduced me to the diverse concepts of adaptation practices in Software, Automobile and Aerospace industries. I fondly remember the support that I received from the Briony Rogers and Christian Urich in the Elwood research group. Roland Lowe and Prof. Karsten Arnjberg Neilsen were very patient, supportive, hosted me in Copenhagen and treated me as their own team member during the development of the Elster creek urban drainage model. Roland and Christian deserve a special mention for dragging me into the sea and making me surf in Australia☺.

I also thank my fellow FRG buddies Kim, Flora, Ngyen, Marilou, William, Jeron, Carlos and Polpot for sharing their knowledge and help. "*Mr. Q*" alias Nguyen Hong Quan has been my motivation and a solid companion in broadening my vision on climate adaptation and to venture into climate adaptation practices in Vietnam.

I am grateful for all the difficult and the almost annoying questions of Dr. Berry Gersonius, my daily supervisor, without which I would have ended up with a big confusing story instead of a thesis. Thanks for asking me the difficult questions, sounding me about the current trends, training me in finding mistakes and also for packing me off to Melbourne for six weeks with a weeks' notice☺.

Prof. Dr. Chris Zevenbergen, my promotor, has always been my mind reader. Notwithstanding my confusing, uncoherent explanations and writings, Chris saw the crux of matter behind it and polished the papers, making me wonder "Dude, how did you do this!!!!". Chris prophesised rightly after my proposal defence that my thesis would be completely different from what I proposed to do. Thank you Chris for the constant encouragement.

Prof. Richard Ashely, Godfather of this thesis on whose feet I submit it, was always available on email and Skype and guided me throughout these four years. I was always looking forward to the weekly skype meetings and the comments on my work. The comments were direct, reflective and with satirical sense of humour. I and my housemates – Tom and Mathew - used to jump like kids, who receive the weekly edition of comics by post, and gather around the screen when his comments arrive. Chris knew where I was going from the beginning but Richard, you were the one guiding and encouraging me all along, asking me not be afraid, to take all the crazy, interesting detours. A thank you is not enough Richard.

Dr. Assela Pathirana, my co-promotor, is the man behind this thesis. I almost gave up the idea of doing PhD nearly five years ago due to my love and passion for working with *Doctors without Borders* in emergency situations. It was the unrelenting pursuit of Assela to secure funding for my research which changed my mind. Most importantly Assela gave me the freedom to chase the crazy ideas I had. Also he had immense trust in me, which played a crucial role in motivating me to complete this research. I owe this PhD to you Assela.

I would be cursed without coffee in my afterlife if I forget to mention Maged, Damon, Daphne, and the lovely team at Café Zondag, from where I wrote this entire thesis. Thank you fellows for the bottomless cups of flat whites, hummus sandwiches and Tajine served with lots of love.

Rev. Walraut and Fr. Avin, my spiritual gurus at Delft, always motivated me to keep going whenever I was on an emotional roller coaster. I thank and remember with gratitude my friends from (i) ISC community especially Mathew, Tom, Punki & Senot, Claudia & Gijs, Claudia & Pavel, Ashish, Arun, Joana, Catarina (Big Cat), Etienne, Mariana & Francesco, Consta & Mithun, Carmen, Ella and Janet; (ii) IHE especially Mamaii Chris, Eva, Tita Marilou, Nadya, Victor, Feranda, Paulo Corgosinho, Sachin & Keerthi, Mohammed & Bahour, Vero, Zara, Pedi, Joesph, Nirajan & Anitha, Noor, Nguyen and Alex; and (iii) all over Netherlands Frederik & Ericka; Assela, Chandini Akka & Kids; Rangarajan Sir & Rani Amma and all other friends. Thank you all so much for being my family at Delft. Dank u wel.

I also thank the next generation of kids: my lovely little niece Manojna (*Ingu pullai*) ☺ and all my lovely little playmates in Delft – Manu, Angila, Benta, Jochem, Katarina, Nicolas, Kora, Sofia, Zara, Ali, Sania – for instilling hope and responsibility in me towards making this world a better place for them.

Also thank you Rajesh Bava for madly following a dream and being a living example for perseverance, determination and courage. I thank my lovely friend Athena and her family, hailing from Normandy in France, for having adopted me as one of their own, showering their love and kindness in all these years. Be it far or near, my friends Kavin, Kirubar, Gowri, Vignes, Esakki Raj, Rakesh, Raja Guru, Murgesan, their families and my aunt Palayamkottai Atthai (Dr.Maragathavalli) have always been my source of strength and refuge.

It is my parent's discipline, blessings and sacrifices that forms the bed rock of my success. I thank Amma and Appa for making me who I am. What will I do without my lovely sister Selvi and an understanding brother in-law Govidasamy? It was their constant support, understanding and shouldering my responsibilities made my research life smooth sailing.

Thank you all ☺ ☺ ☺ .

காலத்தி னாற்செய்த நன்றி சிறிதெனினும்
ஞாலத்தின் மாணப் பெரிது

Verse 110 Tirukural (Thiruvalluar 31 BC)

A helpful act howsoever slight
When timely, acquires true height
Translation by Gandhi (2015)

x

Summary

"பரியது கூர்ங்கோட்டது ஆயினும் யானை
வெரூஉம் புலிதாக் குறின்."

- Verse 599, Tirukural (Thiruvalluar 31 BC)

"Huge bulk of elephant with pointed tusk all armed,
when tiger threatens shrinks away alarmed!"
- Translation *based on* Pope et al. (1886)

Flooding has impacts on human activity. The positive impacts contribute to increased water availability and alluvium which increases agricultural productivity in flood plains. The negative impacts are associated with damage to life, property and productivity. The recent increased damage of flooding compared with the past is due to population growth and accumulation of assets in flood prone areas; more flooding incidents; and, changes in climate. For example, two-thirds of the population of The Netherlands lives in areas that are flood prone from rivers and the sea. According to the Organization for Economic Co-operation and Development, worldwide flooding caused USD 40 billion in losses and affected about 250 million people globally in the year 2015. Many mega-cities such as Bangkok, Jakarta, Paris, Mumbai, New Delhi and New York have been affected by floods in the last decade. Climate change has led to altered precipitation patterns and increase in sea levels, contributing to an increase in flooding. For example, the Australian Government`s Bureau of Meteorology has stated that the sea levels have risen between 2.6 and 2.9 mm every year since 1993 around Australia, and the rise in mean sea level amplifies the effects of high tides and storm surges. The increase in urbanisation, together with the effects of climate change complicate the ways for managing flooding in urban areas. However, there is a range of flood risk management strategies or adaptation measures across countries to mitigate the consequences (e.g., the year 2007 European Union Directive on assessment and management of floods).

Combinations of flood risk management strategies or adaptation measures are seen as the way forward to tackle uncertainties faced by flood risk managers that arise out of climate change and urbanisation, amongst others. Uncertainties cannot be fully resolved. Postponing the decisions under the premise that new insights may emerge and hence reduce uncertainty. The opportunity to increase the resilience of the urban systems in order to minimise negative impacts of uncertainty and maximise positive impacts can be considered as the positive outcome of uncertainty.

Resilience is the ability to: (i) withstand or recover from disturbances; and (ii) anticipate and adapt to change. Resilience towards climate and urban change can be increased by means of flexible adaptation measures. Flexibility in this context can be defined as the presence of opportunities arising from the number of alternative ways to provide services and respond to changing circumstances. As an example, in 2010 the Australian Government released a position paper on adapting to climate change, which considers uncertainty as an opportunity to introduce flexibility and creativity through adaptive measures. Cities such as Melbourne and the State of Victoria have already started including flexibility in their adaptation planning. In contemporary adaptation planning, climate resilience measures are considered together with objectives such as sustainability, productivity and transformations (e.g., transformative adaptation, water sensitive cities). A transformative approach aims to change urban design and supporting structures; the way of living, working and commuting; and the way services are provided in urban areas. Transformation often refers to a system change as opposed to incremental changes, which is the norm in most of the adaptation approaches. Though radical in nature, the transformative approach recommends the use of flexibility and resilience to achieve and sustain the transformation. The concept of 'Water Sensitive City' – i.e. a city being liveable, resilient, sustainable and productive whilst managing all aspects of the water cycle – is gaining popularity among policy makers and planners especially in developed economies. This concept also promotes flexibility as an essential attribute to convert uncertainty into opportunity.

The nature of uncertainties and opportunities prevailing in the financial markets are similar to the uncertainties and risks in urban flood risk management systems. Considering uncertainty as an opportunity to maximise the return on investments is a well-established practice in finance domains. For example, "Options" is an instrument that is used to adjust to fluctuating market prices by engaging in a contract in the present. The contract provides the user the right to buy or sell a product in the future at a price decided in the present. As the option provides a right but not an obligation, the user can decide whether or not to exercise the option based on how the situation unfolds during the course of time. The two pre-requisites for profiting when trading with options are: (i) possessing options - i.e., creating a chance to work with uncertainty; (ii) exercising the option, i.e., converting the option into an opportunity at the time of convenience. Hence, there is scope for converting the (flood) risk into opportunities or favourable outcomes in managing the risk by *a priori* creating two pre-requisites. Firstly, opportunities can be created by pre-planning in the form of pre-defined optional adaptation measures for propitious implementation in future. Secondly, the performance of these measures can be assessed under plausible future scenarios to know when best to implement the measures. This is called managerial flexibility. Adaptation pathways and real options are among the methods that enable managerial flexibility in urban flood risk management.

Adaptation pathways and real options are sequential decision making approaches. They foresee the outcomes of the current and future decisions that might affect the flexibility of a measure or set of measures. An adaptation pathways approach builds flexibility by sequencing the implementation of adaptation measures, so that the urban flood risk management system can adapt to changing climatic, social, economic and environmental conditions. A real options approach values the sequenced adaptation measures in financial terms and helps in determining the sequence of adaptation measures that has the best value for money, which is based on the probability of scenarios. A modified form of the real options approach known as "real in Options" focusses on providing value for the flexibility inherent in having the options in engineered systems such as dikes or drainage systems. This is accomplished by identifying a design configuration that would incur minimum construction, operation, modification and maintenance cost but with the maximum avoided flood damages across a range of scenarios in the future. Real in options is an approach that has been used in the planning and design of large scale infrastructure systems such as high rise buildings, roads and telecommunication network.

The scientific community has tested adaptation pathways and real in options approaches in order to make flood risk and other management strategies flexible in case studies across many countries. Further, these approaches are now finding their way into the planning documents that are addressing flood risk and asset management in the UK and The Netherlands. However, these applications consider only the relationship between the adaption measure and the driver of adaptation such as sea level rise, rainfall or urbanisation. The inter-relationship between the adaptation measures such as measures taken at city level and measures taken at household level are not typically taken into account. Further, the inter-relationships between the adaptation drivers are also not typically considered in the current adaptation planning context. It can be concluded therefore that there is a lack of comprehensive framing of the adaptation responses to take into account the relationships between the adaptation measures and the drivers of adaptation. This lack of comprehensiveness in framing the adaptation responses during planning can also lead to implementation issues. Appropriate structuring of an adaptation response or opportunity in a local context is critical, as this influences the nature and effectiveness of the adaptation. This research focuses on: (i) increasing the knowledge on incorporating flexibility into urban flood risk management systems; (ii) understanding the various aspects of climate and urban adaptation responses; and (iii) development of flexible implementation practices.

The research presented here has developed a generic framework for structuring a multiple perspective approach as a way to increase flexibility in selecting and timing the implementation of adaptation measures. This framework is different from the traditional 'portfolio of measures' approach as it helps to establish the relationship between measures in an adaptation context. The framework has been used to structure the adaptation responses to flood risk in Can Tho city, Vietnam, in order to ascertain the applicability in a practical adaptation context. A context specific adaptation grammar based on 'Systems Engineering' concepts has been used to analyse the structured adaptation responses in Can Tho. The application of context specific adaptation grammar has revealed that a generic framework, such as that presented here, can be used to structure context specific adaptation responses, and it is possible to generate adaptation pathways based on the relationships between the measures.

By demonstrating that flexibility can be enhanced by mapping the relationships between the measures, the scope of the research has been further broadened to create a flexible adaptation planning process. This flexible adaptation planning process identifies where flexibility can be embedded in urban flood risk management systems. The process for this has been developed by drawing on knowledge and procedures used by the automobile and aerospace industries, where flexible adaptation planning is everyday practice. The identification of a flexible water sensitive design component is based on change propagation; i.e. the adaptation measure's ability to minimise or maximise negative and positive impacts in the urban system. This process has been applied here to identify flexible adaptation measures for managing and adapting to flood risk in Elster creek, an urban catchment in Melbourne, Australia. From the application of the process it was found, during the course of this research, that rainwater harvesting systems and flood proofing measures at the household scale are the best measures for incorporating flexibility to tackle the challenges due to flooding; and ensure effective and efficient flood risk management in the future. It was found that the identification of flexible components for urban flood risk management systems based on change propagation can enhance adaptation of cities.

Hence, through this research it has been established that context specific adaptation responses can be structured using a generic framework. However, operational challenges persist in implementing the adaptation measures in managing flood risks, even after identifying flexible components based on change propagation concepts. Addressing the operational and implementation challenges whilst adapting is especially significant in rapidly developing cities where there is a competition for funds between improving the current infrastructure (adaptation) deficits and future adaptation needs. Hence, there is a need to align adaptation actions that lead to improved liveability, sustainability and resilience. The nature of the adaptation responses is similar, at an abstract level, to any other 'complex problem' identified in various domains like software development, manufacturing and supply-chain management. The widely accepted 'agile principles' as identified in these domains are used here for developing equivalent practices in urban adaptation for flood risk management and a set of twelve principles is proposed for operationalising responses in order to adapt to urban flooding. These agile principles have been used to set out four objectives of urban adaptation – flexible incremental measures; common understanding of an adaptation problem; equal importance to adaptation gaps and deficits, i.e., the competition for funds between improving the current infrastructure needs and adaptation to future changes; and stakeholders working together – that can help to translate these principles into tangible outcomes.

This research presented here has focused on increasing the knowledge on incorporating flexibility into urban flood risk management systems by developing and testing: (i) a framework for structuring adaptation responses in a local adaptation context; (ii) a flexible adaptation planning process to identify the flexible adaptation components; and (iii) an agile urban adaptation process to address the operational challenges while implementing flexible adaptation measures. A framework for structuring an adaptation problem in a local adaptation context has been defined, and using this the adaptation problem in Can Tho, Vietnam, have been structured and the adaptation measures that are suitable for the changing context there identified. A process for identifying flexible adaptation components in urban flood risk management systems has been developed and tested in Elster creek in Melbourne, Australia. Incorporating flexibility into adaptation planning and operationalising it through an urban agile process can pave the way for efficient and effective management of urban floods.

Samenvatting

Overstromingen hebben zowel positieve als negatieve gevolgen voor de mens. Positieve gevolgen zijn o.a. een grotere beschikbaarheid van water en sediment, dat de productiviteit van de landbouw in uiterwaarden verhoogt. Negatieve gevolgen zijn verdrinking en schadelijke effecten op gezondheid, eigendom en productiviteit. De recente, toegenomen schade door overstromingen wordt veroorzaakt door bevolkingsgroei en meer economische activiteit in kwetsbare gebieden, frequentere overstromingen en klimaatverandering. Zo leeft twee derde van de Nederlandse bevolking in gebieden die overstromingsgevoelig zijn. Volgens cijfers van de Organisatie van Economische Samenwerking en Ontwikkeling (OESO) veroorzaakten overstromingen in 2015 wereldwijd USD 40 miljard aan schade, waarbij 250 miljoen mensen betrokken waren. Klimaatverandering die door menselijke activiteit is veroorzaakt, heeft voorts geleid tot veranderde neerslagpatronen en versnelde zeespiegelstijging. Gegevens van het Bureau van Meteorologie van de Australische regering geven bijvoorbeeld aan dat de zeespiegel rond Australië sinds 1993 ieder jaar tussen de 2,6 en 2,9 mm is gestegen. De toenemende verstedelijking in combinatie met de effecten van klimaatverandering maakt het moeilijker om overstromingen in stedelijke gebieden te beheersen. Toch hebben verschillende landen en regio's de beschikking over een reeks van strategieën voor de bescherming tegen overstromingen of adaptatiemaatregelen om de gevolgen ervan te beperken.

Dit onderzoek richt zich op hoe om te gaan met de talloze onzekerheden die verbonden zijn aan het ontwikkelen van adaptatiestrategieën en maatregelen voor overstromingen in stedelijke gebieden. Hoofddoel is om de weerbaarheid tegen overstromingen in stedelijke gebieden te vergroten. Dit kan worden bereikt door flexibele adaptatiemaatregelen toe te passen zoals die in dit proefschrift worden besproken. Weerbaarheid (of: veerkracht) wordt gedefinieerd als het vermogen om: (i) verstoringen te weerstaan of zich ervan te herstellen en (ii) op veranderingen te anticiperen en zich eraan aan te passen. Het tweede punt vergt flexibele aanpassingen. Flexibele adaptatiemaatregelen kunnen worden gezien als talrijke pijlen in een pijlkoker, die keuzes mogelijk maken uit een aantal opties, zodat bij een gegeven werkelijkheid uit de meest reactieve optie kan worden gekozen. De hoogwaterkering in de rivier de Thames in het Verenigd Koninkrijk, die grote delen van het achterland beschermt, is zo ontworpen dat de hoogte van de waterkering gemakkelijk in de toekomst kan worden aangepast. Huishoudens in Can Tho, Vietnam, nemen hun toevlucht tot flexibele adaptatiemaatregelen, zoals tijdelijke dijken en het ophogen van woningen. Een belangrijke uitdaging waar dit onderzoek zich op richt is om niet alleen maatregelen te beschouwen die de weerbaarheid vergroten, maar die bovendien de maatschappelijke baten vergroten.

Fundamenteel voor de notie van flexibiliteit is het opstellen van adaptatiepaden in de vorm van mogelijke keuzes waaruit op basis van voortschrijdend inzicht in de omstandigheden kan worden gekozen. Bijvoorbeeld, als het waterpeil van een rivier 30 cm stijgt, dan kan het vloerniveau van woningen langs deze rivier worden verhoogd. Als het waterpeil met 50 cm is gestegen, is het beter om een dijk te bouwen. En als het waterpeil met 70 cm is gestegen, wordt de optie gelicht om zowel het vloerniveau van woningen te verhogen als een dijk te bouwen. Op deze manier vormen adaptatiepaden een reeks van mogelijkheden waaruit op basis van voortschrijdend inzicht met de tijd kan worden gekozen.

Beleidsmakers en planners maken steeds vaker gebruik van adaptatiepaden om strategieën en maatregelen tegen overstromingen flexibel te maken. De kernvraag van dit onderzoek is *hoe* strategieën en maatregelen meer flexibel gemaakt kunnen worden en *waar* de flexibiliteit in het stedelijk watersysteem ingebouwd kan worden.

In dit proefschrift is gebruik gemaakt van kennis uit verschillende disciplines, waaronder methodes uit de informatietechnologie en de auto- en ruimtevaartindustrie en modellering uit de financiële sector. Het onderzoek heeft geleid tot de ontwikkeling van (i) een kader waarbinnen adaptatiemaatregelen vanuit meerdere lagen worden geïntegreerd, (ii) een planningsproces om flexibele adaptatiemaatregelen te identificeren en (iii) een implementatie- en monitoringsproces om strategieën en maatregelen bij te kunnen sturen. Een conclusie van dit onderzoek is dat het vergroten van de veerkracht voor overstromingen mogelijk is door een begrip van klimaatadaptatie in de lokale context (*wat*) en de samenhang tussen adaptatiemaatregelen (*hoe*), het vaststellen van flexibele adaptatiemaatregelen (*waar*) en het operationaliseren van flexibiliteit op een slagvaardige manier (*wanneer*). Als adaptatiemaatregelen worden bekeken vanuit meerdere perspectieven en flexibiliteit wordt meegenomen als een belangrijke eigenschap, dan kunnen de gewenste uitkomsten van klimaatadaptatie worden behaald.

தொகுப்பு

நெகிழ்வான திட்டமிடல் முறைகள் : அதிகரிக்கும் நகரத்தின் வெள்ள பாதிப்புகளை எப்போது எங்கே எப்படி முறைபபடுத்துவது

வெள்ளம் நமது வாழ்வில் முக்கியமான இடத்தை பெற்றுள்ளது. இதன் நேர்மாறையான விளைவுகளாக, அபரிமிதமான தண்ணீர் மற்றும் வண்டல் மண் மூலமாக விளை நிலங்களில் விவசாய சாகுபடி அதிகரிக்கின்றது. எதிர்மறை விளைவுகளாக வாழ்வியல் பாதிப்புகள், உயிரிழப்பு, மற்றும் பொருளாதார (பொருள் இழப்பு மற்றும் தொழில் நிமித்தமான பாதிப்புகள்) இழப்புகள் அடங்கும். இப்போதைய நாட்களில் வெள்ளம் சார்ந்த பாதிப்புகள் பரவலாக அதிகரித்துக்கொண்டே இருக்கிறது. இதற்கு காரணமாக, அதிகரிக்கும் மக்கள் தொகை பெருக்கம் மற்றும் வெள்ள அபாயம் உள்ள பகுதிகளில் உள்ள நிலங்களில் கட்டப்படும் கட்டடங்களையும் கூறலாம். இதனால் வெள்ள சம்பவங்கள் அதிகரிக்கின்றன, மேலும் கால நிலை மாற்றங்களும் ஏற்படுகின்றன. எடுத்துக்காட்டாக நெதர்லாந்தில் மூன்றில் இரண்டு பங்கு மக்கள் வெள்ள அபாயம் உள்ள ஆறுகள் மற்றும் கடல் சார்ந்த பகுதிகளில் வசிக்கின்றனர். Organization for Economic Co-operation and Development யிடம் உள்ள ஆய்வு முடிவுகளின் படி, உலக அளவில் வெள்ளம் 2 லட்சத்து 60ஆயிரம் ரூபாய் மதிப்பிலான இழப்பையும் 25 கோடி மக்களுக்கு பாதிப்பையும் ஏற்படுத்தியுள்ளது.மேலும் மனிதர்களினால் ஏற்படுத்தப்பட்ட காலநிலை மாற்றம் பொதுவாக மழை பொழியும் காலம் மற்றும் முறைகளையும் மாற்றியுள்ளது. அதன் மூலம் கடல் மட்டமும் உயர தொடங்கியுள்ளது. இதுவும் வெள்ளம் ஏற்படுவதற்கு ஒரு முக்கியமான கரணம்.மேலும் ஆஸ்திரேலிய அரசாங்கத்தின் (Bureau of Meteorology) முடிவுகளின் படி ஒவொரு ஆண்டும் ஆஸ்திரேலியா சார்ந்த பகுதிகளின் கடல் மட்டம் 2.6mm முதல் 2.9mm வரை அதிகரித்துக்கொண்டுள்ளது. இந்த அதிகரிக்கும் கடல் மட்டம் கடல் அலைகளின் உயரமும் , புயல் பாதிப்புகளையும் பெருக்குகிறது. நகரமயமாக்கலும், காலநிலை மாற்றமும் சேர்ந்து வெள்ள பதிப்புகளில் இருந்து நகரங்களை காப்பாற்றுவதில் இருந்து மேலும் சிக்கலாக்குகின்றன. எனினும், பரவலான பேரிடர் மேலாண்மை உத்திகளும் மற்றும் செயல்பாட்டு முறைகளும் முன்னுதாரணங்களுடன் எல்லா நாடுகளிடமும் செயல்பாட்டில் இருக்கின்றன.

இந்த ஆராய்ச்சி, நகரம் சார்ந்த வெள்ள பாதிப்புகளுக்கான நிச்சயமற்ற தன்மைகள், எதிர்கொள்ளும் சவால்கள் மற்றும் திட்டமிடும் முறைகள் குறித்து அலசுகிறது. நம்மிடையே உள்ள அறிவியல் சார்ந்த தகவல்கள், அதன் புரிதல்கள் மற்றும் தகவல் பரிமாற்றம் குறித்த எதிர்பார்ப்புகள் ஆகியன மேலும் இந்த ஆராய்ச்சிக்கான சவால்களை அதிகரிக்கின்றன. நமது ஆராய்ச்சியின் முதல் நோக்கம் நகரங்களில் ஏற்படும் வெள்ள பாதிப்புகளை தடுப்பதாகும். இதை நெகிழ்வான திட்டமிடல் முறைகளினால் ஆராய்ந்து செயல் படுத்த முடியும். பொதுவாக மீட்டெழும் தன்மை என்பது எந்த விதமான இடையூறிலிருந்தும் தாக்கு பிடிக்க கூடிய அல்லது பாதிப்பிற்கு பின்னும் பழைய நிலையை திரும்ப பெற்று எதிர்காலத்தில் எல்லா விதமான பாதிப்பிற்கும் சவால்களை எதிர்பார்த்திருப்பதாகும். நெகிழ்வான திட்டமிடல் அபரிமிதமான வழி முறைகள் கொண்ட செயல் முறையாகும். இந்த ஒவ்வொரு வழி முறையும் பல்வேறு மாற்று வழிகள் / முறைகளில் இருந்து தெரிவு செய்யப்பட்ட சாதகமான பலன்களை அளிக்கக்கூடிய வழிமுறையாகும். இங்கிலாந்தின் தேம்ஸ் நதிக்கரையில் அமைக்கப்பட்ட தடுப்பு அரண் எதிர்காலத்தில் அதிகரிக்கக்கூடிய வெள்ள பாதிப்புகளுக்கு ஏற்ப தடுப்பு அரணை உயர்த்திக்கொள்ள கூடியதாக அமைந்துள்ளது. அனால் வியட்நாம் நாட்டில் உள்ள கேன் தொ நகரத்தில் கட்டப்பட்ட கட்டிடங்கள் தற்காலிக தடுப்பு அரண் மற்றும் மாறும் தரைமட்டம் / அடித்தளத்துடன் அமைக்கப்பட்டதாகும். இந்த வகை திட்டமிடல்களில் மிக முக்கியமான முடிவெடுக்கும் சக்தியாக, முறையான திட்டமிடல் முறைகளுடன் அனுமதிக்கப்பட்ட நிதி மூலதனமும் அதன் மூலம் திரும்ப கிடைக்கப்பெறும் நிதி ஆதாரமும் விளங்குகிறது.

இணக்கமான / நெகிழ்வான தேர்ந்தெடுத்தல் முறை என்பது, எல்லா வகைகளிலான திட்டமிடல் முறைகளை ஆராய்ந்து, சூழ்நிலைக்கேற்ப சரியான பாதையை அனுமானிப்பதாகும். எடுத்துக்காட்டாக, ஆற்றின் நீர்மட்டம் 30cm உயரும்போது எல்லா கட்டடங்களின் தரைமட்டமும் தற்போதுள்ள ஆற்றின் நீர்மட்டத்தில் இருந்து 30cm உயர்த்தி கட்ட வேண்டும். மேலும் 20 வருடங்களில் ஆற்றின் நீர்மட்டம் 50cm உயரும்போது தடுப்பு அரண் அமைத்துக்கொள்வது சரியான தேர்வாகும், அதேபோல் ஆற்றின் நீர்மட்டம் 70cm உயரும்போது கட்டடங்களின் தரைமட்ட உயரத்தை உயர்த்திக்கொள்வதுடன் தடுப்பு அரண் அமைப்பதுவும் சரியான தெரிவாகும். இதுபோல், எல்லா வழி முறைகளையும் கண்டறிந்து,

பின் அதிலிருந்து காலம் மற்றும் சூழ்நிலைகளுக்கு ஏற்ப நம் சரியான வழிமுறையை தேர்ந்தெடுக்க வேண்டும்.

இன்றைய நகர வடிவமைப்பாளர்கள் வெள்ள தடுப்பு முறைகளுக்கு வலுக்கூட்ட வெவ்வேறு வகையான மாற்றியமைக்கப்பட்ட செயல் முறைகளை நாடுகிறார்கள். இந்த விரிவான அணுகுமுறைக்கு வலு சேர்ப்பதே இந்த ஆராய்ச்சியின் நோக்கமாகும், எனவே இந்த விரிவான ஆராய்ச்சியின் முடிவில், i) உள்ளூர் சூழலுக்கேற்ப தகவமைத்துக்கொள்ளுதல் (என்ன செய்ய வேண்டும்), ii) ஒவ்வொரு தகவமைப்பு முறைகளுக்குமான தொடர்புகள் (எப்படி செய்ய வேண்டும்) iii) இணக்கமான தகவமைப்பு முறைகள் (எவ்வாறு செய்ய வேண்டும்), iv) சுறுசுப்பான / விரைவான செயல்பாட்டுக்கு இணக்கமான முறைகள் (எங்கே செய்ய வேண்டும்) போன்றவை நகர வெள்ள தடுப்பு செயலாக்கத்திற்கு / பணிகளுக்கு வலுகூட்டுவதாகும்.

இந்த விரிவான அணுகுமுறையின் விளைவாக, இந்த ஆராய்ச்சியில் தகவல் தொடர்பு, ஆட்டோமொபைல், விண்வெளி ஆராய்ச்சி மற்றும் உற்பத்தி தொடர்பான செயல்பாட்டு முறைகளும் மற்றும் நிதி துறை சார்ந்த செயல் மாதிரிகளும் உபயோகப்படுத்தப்பட்டுள்ளன. இந்த ஆராய்ச்சி முடிவுகள் i) கட்டமைப்பு தொடர்பான முன்னேற்பாடான செயல் முறைகள் ii) இணக்கமான திட்டமிடல் மூலம் கண்டறியப்படும் சிறந்த செயல் முறைகள் மற்றும் iii) லாகவமான செயல்முறைகள் மூலமாக சிறிது சிறிதாக மாறிவரும் தொடர் கண்காணிப்பு முறைகளின் மூலமாக மேம்படுத்தப்படுகின்றன. மேலும் இணக்கமான திட்டமிடல் மற்றும் பல்வேறு செயல்பாட்டு முறைகள் தொடர்பான ஆராய்ச்சிகள் வியட்நாம் நாட்டில் உள்ள கேன் தொ மற்றும் ஆஸ்திரேலியா நாட்டில் உள்ள மெல்போர்ன் போன்ற நகரங்களில் செயல்படுத்தப்பட்டுக்கொண்டிருகிறது.

Table of Contents

1 Introduction

வான்நின்று உலகம் வழங்கி வருதலால்
தானமிழ்தம் என்றுணரப் பற்று

- Verse 11, Tirukural (Thiruvalluar 31 BC)

"The world its course maintains through life that rain unfailing gives;
Thus rain is known as the true ambrosial food of all that lives"
- Translation *based on* Pope et al. (1886)

Water is essential for sustenance. When there is either an insufficiency or excess of water, all life and functions based on water is affected. Flooding is the undesirable consequences of there being too much water in the wrong place. Flooding has positive and negative impacts on human activity. The flourishing civilisations in flood plains attests to the positive effects of flooding. For example, many ancient Egyptian civilisations which thrived on the banks of River Nile were dependent on the annual floods for fertility of agricultural land among other benefits (Hassan 1997). However, floods have the potential to cause fatalities, displacement of people and damage to the environment and property. This can severely compromise the liveability and economic development of a community and damage ecosystems. Climate change has led to altered precipitation patterns and increased sea levels, contributing to an increase in flooding (IPCC 2014a). The increase in urbanisation, together with the effects of climate change complicate flood management in urban areas (Revi et al. 2014).

The magnitude and urgency of the need to adapt to climate change are such that addressing it has been taken up as one of the sustainable development goals - Goal 13 (SDG13) by the United Nations (UN 2015). SDG13 emphasises the need for strengthening resilience and adaptive capacity to climate related hazards and natural disasters and also calls for the integration of climate change measures into policies, strategies and planning. Adapting to urban floods is one of the major needs of climate adaptation, where integration of climate change responses into flood risk management policies, strategies and planning at international, national, regional and local levels is now the norm. However, much of this integration lacks effectiveness or real commitment (Anguelovski et al. 2014; Ashley et al. 2007; Deltacommissaris 2014; EU 2007; MDP 2013; Revi et al. 2014).

Since time immemorial, people have been tackling flooding either by protecting themselves from floods or by living with floods. The Netherlands is seen by many as providing an example of pioneering flood protection using dikes and an efficient overland drainage system comprising a network of canals and pumps (Deltacommissaris 2014). In the Vietnamese Mekong Delta, people have adapted to floods and are living with floods. Flooding is a common seasonal phenomenon in the Mekong Delta and as well as being threatening, also brings benefits as the flood water makes the agricultural areas more fertile by bringing nutrient rich alluvium (Wesselink et al. 2016). Most of the houses in the Mekong delta have been adapted to flooding, with houses being either built elevated on stilts to avoid inundation, or protected with seasonal measures such as temporary dikes, put in place in rural areas (Birkmann et al. 2012). In the urban areas such as Can Tho, the houses are made wet proof, such as waterproof flooring and walls, easily movable furniture, electrical sockets at windowsill level, etc., to minimise the damage due to standing water (DWF 2011). It is also possible to combine the strategies which rely on flood protection, i.e., avoiding flooding, and strategies that adapt flooding, i.e., living with floods.

Haasnoot et al. (2013), Sayers et al. (2013), Gersonius et al. (2016) recommend combining strategies to manage urban flood risk management such as: (i) retaining flood waters in upper catchments; (ii) relieving flood impacts using diversionary arrangements; (iii) resisting floods using dams and embankments; (iv) retreating from flood prone areas; (v) adapting to floods such as wet proofing or making water an integral part of the urban landscape; and (vi) preparing for the consequences of flooding in case of such an eventuality. Cities such Rotterdam, Copenhagen, London and Melbourne have already started combining strategies in planning and implementing adaptation measures (City of Melbourne 2016; EEA 2016; HM Government 2016).

Planning of water management in general has, for much of the twentieth century, been based on the presumed stationarity of climatic variables. However, the obsolescence of stationarity in hydrological systems is increasingly being recognised in the scientific community (Milly et al. 2008). Further there are uncertainties regarding adaptation that are associated with socio-economic scenarios, climate models, biophysical impact models, vulnerability assessments and appraisal of adaptation measures (Street and Nilsson 2014). It has been assumed by engineers and planners that adding a simple margin of safety to the historic record (which is assumed to be stationary without much error) is enough to take into account any and all future uncertainties (e.g. additional freeboard to dikes and critical infrastructure to take care of uncertainties due to the anticipated rise in sea levels in areas like the Mekong Delta (SCE 2013), and as set out the UK's Flood resilience strategy (HM Government 2016)). However, current developments on uncertainties related to climate (e.g., Milly et al. (2008)) and other societal factors (e.g., O'Neill et al. (2015)) have to be taken into account more rigorously in planning and implementing practical water management measures. In addition to advances in knowledge, a strategic shift in planning and implementation can be noticed in the way the various levels of Government tackle the uncertainties that are associated with climate change. For example, the Australian Commonwealth Government considers uncertainty as an opportunity to introduce flexibility and creativity while adapting to changing climate (Commonwealth of Australia 2010). Flexibility is seen as a desirable attribute that enhances system capabilities and functionality in the face of uncertainty (Schulz et al. 2000).

Flexibility, according to *The New Oxford dictionary of English (First edition)* is defined as the ability to be easily modified to respond to altered circumstances or ready and able to change so as to adapt to different circumstances (Pearsall 1998). The ability to keep some options open so as to provide opportunities for the decision maker to take action after uncertainties are revealed, is defined as flexibility in the context of uncertainty (De Neufville and Scholtes 2011). Flexibility entails changes in structure, scale, functionality and operational objectives as the result of external intervention by a change agent (Fricke and Schulz 2005; Ross et al. 2008). All forms of flexibilities- flexibility to defer, flexibility for expansion or contraction, flexibility to switch - should be explored for adapting under favourable / unfavourable conditions. The search for appropriate forms of flexibility could be extended to the whole range of measures available across the adaptation strategies to explore the possibilities of keeping some of the measures open, allowing change to their scale and functionality. Radhakrishnan et al. (2016) classify the various forms of flexibility, which are prevalent in urban flood risk management as follows: (i) structural flexibility (e.g., TE2100 (2012)); (ii) managerial flexibility (e.g., Haasnoot et al. (2012b); Gersonius et al. (2013); Woodward et al. (2014)); (iii) functional flexibility (e.g., Balmforth et al. (2006); EEA (2016)) ; (iv) operational flexibility (e.g., Rickard (2009)); and (v) strategic flexibility (e.g., Zevenbergen et al. (2015a)). These are considered further in the following.

Structural flexibility is the possibility to incorporate flexibility into structures that facilitates the deferral of expansion, such as the heightening of dikes and upgrading of drainage systems (Gersonius et al. 2013; Woodward et al. 2014). The Thames Estuary action plan (TE2100) for example recommends incorporating structural flexibility in the form of strong foundations constructed during the replacement of existing gates of the Thames Barrier so that they can take the additional load if future raising of crest levels is needed (TE2100 2012). Enhancing flood protection by means of employing movable, demountable or temporary flood protection systems can be a means of bringing operational flexibility. An example of movable flood protection is the Maeslantkering storm surge barrier in The Netherlands[1] that enables navigation as well as protection against sea surges. The demountable flood defences in Shrewsbury and Bewdley along the River Severn in the UK[2] ensures flood protection during flooding and access to river banks when there is no flooding (Rickard 2009).

Functional flexibility of a component or a system may be defined as an attribute that enables it to perform a task which is not usually expected of it during normal operating conditions or at a particular point in time. For example a road could be designed to convey excess surface flow and a park could also be used as a detention basin (Balmforth et al. 2006), whilst providing their main functions as a transport route and a recreational area respectively. Strategic flexibility or objective flexibility may be defined as planning and executing a set of measures that are based upon a particular strategy that does not prevent or creates only minimal hindrance when switching to a completely different strategy (e.g. Delta plan van Haegen and Wieriks (2015)). Managerial flexibility is the ability to keep the adaptation measures open for future adaptation or postponing adaptation measures until the time when the cost of further delay would be more than the benefits of doing so (Radhakrishnan et al. 2016). An example for managerial flexibility is the design of water supply systems to produce drinking water through various treatment processes in Singapore (Zhang and Babovic 2012).

Managerial flexibility may be incorporated into urban flood risk management using sequential decision making approaches such as adaptation pathways and real options (Simpson et al.

---

1 http://www.keringhuis.nl/index.php?id=13 [accessed May 2017]

2 http://evidence.environment-agency.gov.uk/FCERM/en/FluvialDesignGuide/Chapter9.aspx?pagenum=10

[accessed August 2016]

2016). An adaptation pathway approach builds flexibility into decision making processes by sequencing a set of adaptation measures based on a 'tipping point' to changing circumstances in a range of plausible future conditions (Haasnoot et al. 2012b). Tipping points are the points in time in future or predetermined values of variables such as sea level rise, rainfall, at which the objective of an adaptation strategy is no longer met or the functionality of an adaptation measure is not valid (Kwadijk et al. 2010). A real options approach (Dixit and Pindyck 1994; Myers 1984) values the sequenced adaptation measures in financial terms and helps in determining the sequence of adaptation measures that has the best value for money, based on the probability of scenarios. A real option approach comprises the multiple representations of decisions that are taken as a response to changes in circumstances and the probabilities associated with the various changes in circumstances. Changes in circumstances – such as demand, rainfall intensity and sea level rise – trigger a series of decisive actions. These changes create independent decision pathways and the cost and benefits along these pathways can be determined (Zhang and Babovic 2012). There are also approaches that combine real options and adaptation pathways in selecting the preferred pathways over a range of pathways (Manocha and Babovic 2016).

The adaptation pathways, real options and real in option approaches are now being used in climate adaptation planning such as in the UK (e.g., defra (2009)) and in The Netherlands (e.g., Delta commissaris (2014)). These flexibility incorporation and evaluation approaches are effective in sequencing adaptation measures when the adaptation context is clear and the relationships between the adaptation measures and drivers are direct and simple. However, more typically when addressing flood risks there may be multiple uncertainties arising from many sources and multiple contexts or hazards (Simpson et al. 2016). For example, adaptation in a particular context might face uncertainties due to simultaneous increases in sea level *and* rainfall. The adaptation becomes more complicated when the adaptation is also facing multiple uncertainties in multiple contexts such as climate change (e.g., IPCC (2014a)), socio-economic change (e.g., O'Neill et al. (2015)) and political change (e.g., O'Brien (2015)). In such circumstances, understanding the adaptation context and framing the adaptation 'problem' becomes a crucial step in the planning and implementation of the adaptation measures.

The incremental and sequential adaptation approaches such as adaptation pathways, real options and real-in-options are useful in understanding and evaluating the overall flexibility in implementing adaptation measures. Real options and real-in-options approaches are useful in evaluating the performance of adaptation measures under uncertainty in monetary terms, whereas adaptation pathways are used to assess a set of pathways based on the tipping points (e.g., Gersonius et al. (2013); Haasnoot et al. (2012b); Woodward et al. (2014)). There is a change in the way the adaptation measures are assessed, which have been mostly based on singular objectives, such as economic efficiency or resilience, towards assessments based on multiple objectives. Also, urban adaptation is increasingly seen as an essential component for improving the quality of life and wellbeing in urban areas (City of Melbourne 2016; EEA 2016; Kleinert and Horton 2016; Sallis et al. In press). Hence, the incremental and sequential adaptation approaches have to support the transformative urban adaptation approaches, (EEA 2016; Revi et al. 2014).

A transformative approach aims to change the urban design and structures; the way of living, working and commuting; and the way services are provided in urban areas (EEA 2016; Revi et al. 2014). Transformational adaptation is the way of using behaviour and technology to change biophysical, social or economic components of a system fundamentally (EEA 2016). Transformation often refers to a system change as opposed to incremental changes, which is the norm in most of the adaptation approaches. Though radical in nature, the transformative approach recommends the use of flexibility and resilience to achieve and sustain the transformation. For example, the concept of Water Sensitive City, i.e. a city being liveable, resilient, sustainable and productive, whilst managing all aspects of the water cycle, is gaining popularity around the world in developed economies (Ferguson et al. 2013b; Howe and Mitchell 2011; Wong and Brown 2009; Wong 2006). Australian cities such as Melbourne and the State of Victoria have already started to include flexibility in their adaptation planning (City of Melbourne 2016; Victoria 2016a; Victoria 2016b). Further, flexibility is also seen as an essential character of urban planning and infrastructure to deal with transformation – in enabling a smooth transition – in objectives such as becoming a water sensitive city from the current transition state of water supply city, waterways city, etc., (Ashley et al. 2013b; Brown et al. 2009; Howe and Mitchell 2011).

A prerequisite for realizing the potential of flexibility is to understand the adaptation context. Framing an adaptation 'problem' or 'responses' in a given context is critical, as this influences the nature and effectiveness of adaptation. There are many ways to analytically frame adaptation (e.g., UNEP (2014)). Particular contextual assumptions, methods, interpretation, and values that different stakeholders bring is central to planning adaptation pathways in an urban context. According to Wise et al. (2014) a broad conceptualisation of adaptation based on adaptation pathways is essential. Conceptualisation of adaptation should comprise the consideration of: (i) the implications of path dependency; (ii) any interactions between adaptation plans; (iii) global changes, such as climate change and socio-economic changes; and (iv) situations where values, interests and institutions constrain societal responses to change. Such a conceptualisation can also reduce the barriers during the implementation of adaptation measures.

Implementation of adaptation measures is complicated and needs a thorough understanding of the context, characteristics of adaptation measures, the roles and responsibility of various stakeholders in the local adaptation context (Ellen et al. 2014; Kleinert and Horton 2016; Phi et al. 2015). Framing of adaptation responses and proactive analysis of implementation barriers will ease the implementation and operationalisation of adaptation measures. However, operationalising flexibility - i.e., implementing short and long term flexible adaptation measures with a long term perspective - is a challenge, as short term measures are usually preferred by influential decision makers such as politicians (Buuren et al. 2013; Edelenbos 2005). Planning and implementing flexible short team measures, though done with in the ambit of long term change, in some context can be misunderstood as indecisiveness and can lead to loss of credibility and legitimacy (Buuren et al. 2013).

Framing of adaptation responses and predicting the outcomes of adaptation measures is difficult in cities as they comprise a myriad of physical, social, environmental, economic and political systems that interact, self-organize and produce emergent change (Dunn et al. 2016). In this context cities may be considered as "complex adaptive systems" based on the definition of the term by Holland (1992), i.e., *"systems that reorganise and change their components to adapt themselves to the problems posed by their surroundings"*. Alternatively, cities could also be considered as "complex adaptable systems" by combining the "adaptable system" definition of Oppermann (1994) and the "complex system" definition of Cilliers (2001). Adaptable systems, are systems whose components can be changed by the decision maker or society in order to adapt to the changing circumstances (Oppermann 1994). Complex systems are defined as systems that cannot be explained, described or predicted with any degree of accuracy (Cilliers 2001).

The interactions between the various systems in a city, by considering the city as a complex adaptable system, may be studied using "system approaches" such as outlined by Fratini et al. (2012), who focus on the inter-relationships between the natural and technical systems in urban areas. Risk-centered, systems approaches can facilitate understanding of the complex interactions and dependencies across environmental, social, and human systems (Ebi et al. 2016). This understanding based on a systems approach will reduce the risk of maladaptation, i.e., the possibility that adaptation might increase the vulnerability of other stakeholders or sectors in the future (Barnett and O'Neill 2010). Hence, ascertaining and understanding the characteristic features of adaptation measures based on a 'complex adaptable systems approach' is essential in a multiple adaptation context in order to incorporate flexibility.

The aim of the research is to enhance the effectiveness of flexible adaptation practices towards achieving the desired adaptation outcomes in an urban environment. Hence this research focuses on: (i) increasing knowledge on incorporating flexibility into urban flood risk management systems; (ii) understanding the various aspects of climate and urban adaptation responses that are related to urban flooding; and (iii) development of flexible implementation practices towards adapting to urban flooding.

Based on the narrative explained in this chapter and the gaps identified in the urban climate adaptation context – especially in the context of flooding – within which the research is set, the following research questions have been formulated:

1. How to structure a climate adaptation problem in a local context?
2. Upon structuring a climate adaptation problem, how to prioritise a set of adaptation measures in a specific local context?
3. How to embed flexibility into a local adaptation context?
4. How to operationalise flexibility in an adaptation context?

This thesis is structured as follows:

With a focus on flooding: *Chapter 1* (this chapter) sets the context for the research and how and why the various classifications of adaptation responses can be grouped for the specific needs such as integrating responses, assess the effectiveness of responses and implementation of responses. *Chapter 2* considers and aims to understand the climate adaptation problem in a local context using a multiple perspective framework for structuring the climate adaptation problem. *Chapter 3* establishes the relationships between adaptation measures for enhancing flexibility. *Chapter 4* demonstrates how in a local adaptation context – using as an example Can Tho, Vietnam – adaptation pathways can be generated and preferred adaptation pathways selected using a context specific adaptation grammar approach. *Chapter 5* presents a flexible adaptation planning process (WSCapp) which has been developed in this research by utilising an approach to incorporating flexible processes that is prevalent in the automobile and aerospace industry. *Chapter 6* demonstrates the application of the flexible adaptation planning process in an urban flood adaptation context in Melbourne, Australia. *Chapter 7* elucidates the "Agile urban adaptation planning process", which is based on the practices from the software industry for operationalising flexibility in an ever changing urban context. Finally, *Chapter 8*, provides recommendations and sets out the limitations of this thesis, with scope for future research.

The answers to the research questions are sought through (i) understanding of relationships between the adaptation responses in a local context; (ii) assessment of the effectiveness of adaptation responses in achieving the multiple objectives; and (iii) establishing the processes which facilitate the implementation of these responses. Understanding the basic features and characteristics of adaptation measures are essential to arrive at the answer to the research questions. One of the ways to understand the characteristics is to study how adaptation responses are classified. There are various classification of adaptation responses (Haque et al. 2014; Jabeen et al. 2010; Milman and Warner 2016; Mycoo 2014; Næss et al. 2005; Pathirana et al. 2017; Schaer 2015; Thorn et al. 2015; Walker et al. 2013; Wamsler and Brink 2014b) such as (i) actors (public, private); (ii) scale (household, institutional); (iii) approach (autonomous, planned); (iv) medium (physical social, economic); (v) type (current deficits, future requirements); (vi) purpose (prevention, response, recovery); and (iv) time span (short term, long term). The various classifications of adaptation responses are illustrated in Figure 1-1.

From Figure 1-1 it can be seen that the seven classification of measures have been categorised based on who is adapting, when they are adapting, how are they adapting and what they are adapting to. Also this classification enables the decision maker or researcher to contextualise a particular adaptation measure. For example, elevation of floor levels in houses is usually an autonomous measure implemented with the resource of household or individual can also be a planned public measure if it is supported by the government through vulnerability reduction programmes.

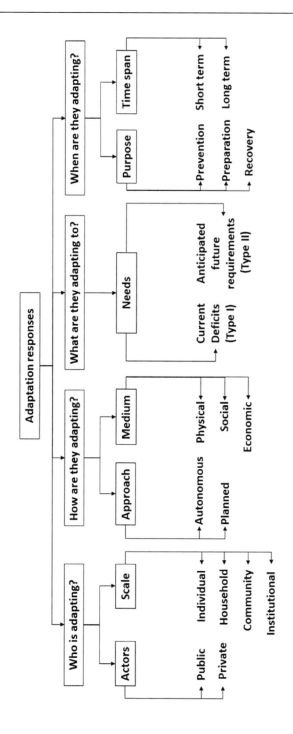

Figure 1-1 Classification of adaptation responses

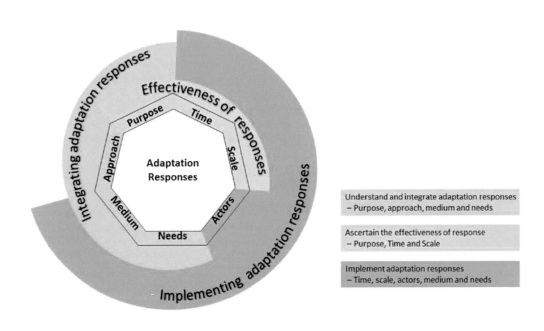

Figure 1-2 Need or puropse based grouping or reclassification of adaptation measures

Also this thesis is based on the premise that a grouping or reclassification of adaptation responses is essential to satisfy the specific need (purpose) at every phase of adaptation (Figure 1-2).

1. The purpose, nature, form and type of adaptation responses determine the extent to which adaptation responses can be structured and integrated at the adaptation planning stage (Chapter 2, 3 and 4).

2. The purpose, time and scale of the adaptation measures determine the effectiveness of adaptation responses towards achieving the adaptation objectives (Chapters 5 and 6).

3. The time, scale, actors, form and type of the adaptation measures determine the ease of implementation, continuous stakeholder participation and joint learning as and when the adaptation measures evolve (Chapter 7).

In Chapter 1, the theory behind the following adaptation planning processes such as Adaptation tipping points (Kwadijk et al. 2010), Adaptation pathways (Haasnoot et al. 2012b) and Real-in-options (Gersonius et al. 2013; Woodward et al. 2014), and their application are also explored in detailed. Further this chapter set the research context through various perspectives for adaptation such as engineering, planning, social, political classification of adaptation measures such as short term vs long term, autonomous vs planned; reactive or anticipatory (e.g. Walker et al. (2013)). The anticipated outcome of this chapter is the understanding of the development of these the adaptation planning process; their strength and weakness; context in which they are applied.

The identification of gaps in the adaptation planning process in understanding the adaptation context and implementation of adaptation measures is important, as contemporary adaptation contexts are multiple adaptation contexts. The adaptation is driven by multiple drivers such as climate, urbanisation and political drivers. Understanding the adaptation context and framing the adaptation 'problem' becomes a crucial step in the planning and implementation of the adaptation measures. Chapter 2 and 3 sets out to review the various adaptation planning perspectives, drivers and objectives. The identification of gaps in knowledge and practice which are found out upon the review of adaptation planning process and is expected to contribute towards the creation and application of an adaptation framework to structure and integrate adaptation responses.

Chapter 4 is about resolving the structured adaptation planning problem and the integrated adaptation responses. In a multiple adaptation context there are number of adaptation measures and drivers. There are inter-relationships among the measures, drives and between the drivers and the measures. Establishing these relationships and assessing the effectiveness of adaptation measures in a complex adaptation system across a range of scenarios is a challenging task. Chapter 4 elaborates on a context specific adaptation grammar which can accommodate the context specific adaptation relationships in a local adaptation context can be used to resolve the structured adaptation responses (Islam 2016; Radhakrishnan et al. Submitted-a).

The adaptation planning process such as adaptation pathways approach deals with the question of how adaptation can be made flexibility, whereas real option and real-in-options ascertain the value of flexibility in monetary terms. These measures do not ascertain the place, where flexibility can be incorporated. Chapter 5 elaborates on a flexible adaptation planning process (WSCapp) which identifies the adaptation response or the urban system where flexibility can be embedded. WSCass is created based on the flexible design practices that are prevalent in automobile and aerospace sector such as Suh et al. (2007) and Eckert et al. (2004). Chapter 6 is about the application of WSCapp in an urban flood adaptation context in Melbourne, Australia.

Operational challenges persist in implementing the adaptation measures in managing flood risks, even after identifying flexible adaptation responses based on change propagation concepts. Addressing the operational and implementation challenges whilst adapting is especially significant in rapidly developing cities where there is a competition for funds between improving the current infrastructure (adaptation) deficits and future adaptation needs. Hence, there is a need to align adaptation actions that lead to improved liveability, sustainability and resilience. Chapter 7 elaborates on the development and application of agile urban adaptation process which aims are identifying the agile urban agile components that would facilitate implementation and monitoring of continuously evolving responses, enable stakeholder involvement and joint learning. Agile adaptation practices, i.e., quickly adapting to changes is a common practice in automobile manufacturing sector, fast moving consumer goods and software development (Beck et al. 2001; Fowler and Highsmith 2001; Meredith and Francis 2000; Zhang and Sharifi 2000).

The anticipated outcome of this thesis is the inclusion or enhancement of flexible adaptation planning processes in the practices of city managers and researchers in deciding when, where and how to include flexibility for increasing urban flood resilience.

2 Structuring climate adaptation through multiple perspectives: Framework and case study on flood risk management

"எப்பொருள் யார்யார்வாய்க் கேட்பினும் அப்பொருள்
மெய்ப்பொருள் காண்ப தறிவு"
   Verse 423, (Thiruvalluar 31 BC)

"Though things diverse from diverse sages' lips we learn,
'Tis wisdom's part in each the true thing to discern"
Transation based on Pope et al. (1886)

Adaptation to climate change is being addressed in many domains. This means that there are multiple perspectives on adaptation; often with differing visions resulting in disconnected responses and outcomes. Combining singular perspectives into coherent, combined perspectives that include multiple needs and visions can help to deepen the understanding of various aspects of adaptation and provide more effective responses. Such combinations of perspectives can help to increase the range and variety of adaptation measures available for implementation or avoid maladaptation compared with adaptations derived from a singular perspective. The objective of this chapter is to present and demonstrate a framework for structuring the local adaptation responses using the inputs from multiple perspectives. This framework was demonstrated with reference to the management of flood risks in a case study Can Tho, Vietnam. The results from the case study show that framing of adaptation responses from multiple perspectives can enhance the understanding of adaptation measures, thereby helping to bring about more flexible implementation practices.

This chapter is based *on the journal article "Radhakrishnan M, Pathirana A, Ashley R, Zevenbergen C (2017) Structuring climate adaptation through multiple perspectives: Framework and case study on flood risk management Water 9: 129"*

2.1 Introduction

The magnitude and urgency of the need to adapt to climate change is such that combating it has been taken up as one of the sustainable development goals - Goal 13 (SDG13) by the United Nations (UN 2015). In addition to this urgency, the uncertainties related to climate and other societal factors have to be taken into account more rigorously while planning and implementing practical water management measures (Street and Nilsson 2014). The most common approach to adaptation planning is based on 'singular perspectives' such as looking at the issues from either technical, political or other perspectives, or in often separated sectors such as urban development, drainage, transport or public health (Vink et al. 2014). There are concerns that the individual adaptation strategies arising from these differing perspectives might lead to maladaptation; i.e., increase the vulnerability of other sectors, systems or groups, or lead to inefficiencies in implementation (Barnett and O'Neill 2010). In each case there is a need to assess the risk of maladaptation ideally from the beginning and throughout the adaptation planning process (Magnan et al. 2016).

The outcomes of adaptation planning generally reflect the perspective (lens) through which the adaptation challenges have been analysed. This is because there are defined frameworks, i.e. the basic structures or underlying concepts, which guide response assessment in particular domains and hence the nature of the outcomes. Frameworks tend to 'narrow down' an issue using a particular perspective to structure the adaptation challenge - i.e. arranging it according to the concept, identifying patterns to account for, or sequencing accordingly - in order to get the outcome desirous by the perspective. From a social perspective for example, this involves enhancing or maintaining the liveability of part of an urban area.

There are recent integrated adaptive policy planning mechanisms such as dynamic adaptive policy pathways (Haasnoot et al. 2013) that bring flexibility into the perspective, accommodate multiple perspectives and are also geared for unexpected feedback mechanisms. However, these planning mechanisms are difficult to implement as prevailing practices are not sufficiently flexible (Ellen et al. 2014). Dupuis and Knoepfel (2013) in reviewing the climate change adaptation plans and their subsequent implementation in Switzerland and India, state that, in-spite of the decision makers' desire for framing adaptation policies, they encounter difficulties in implementing adaptation practices. In The Netherlands, adaptation based on flexibility and learning - although appreciated from a planning perspective – is seen to be less credible and without clarity by the general public and many stakeholders, resulting in resistance to implementation (Buuren et al. 2013). Regardless of the adaptive capacity of the country, climate change adaptation has serious implementation issues (Buuren et al. 2013; Dupuis and Knoepfel 2013). Lack of managerial flexibility in implementation is also often attributed to the "command and control" nature of the policies that govern the implementation of adaptation measures (Ellen et al. 2014) and can also be the result of protectionism, or failures to cooperate, between constituent parts of the same organization (Cettner et al. 2014b).

The objective of this chapter is to present a framework for structuring the local adaptation responses using the inputs from multiple perspectives. The resulting comprehensive approach will help to better understand and bridge the gap between climate adaptation planning and adaptation implementation in an urban environment that is currently adapting to climate change, focusing on flood risks

The approach taken in this chapter is to provide a means to enhance the managerial flexibility to adapt and to reduce the risk of maladaptation, by structuring the adaptation challenges through multiple perspectives, and by looking at the link between the adaptation measures and parameters across perspectives. We understand and define parameters as external drivers such as sea level rise, rainfall acting on the system and endogenous factors such as poverty or household incomes within the system. Enhancement of managerial flexibility is achieved by implementing measures in the immediate term or longer into the future, or the implementation of other measures that can lower the risk of over or under investment. While the quantification and identification of flexibility in contributing to adapting to future events is still the subject of much ongoing research (Maurer 2013), the approach set out here contributes by explicitly addressing the means of identification and of increasing flexibility to respond among the various optional adaptation measures.

The chapter is organised as follows: (i) an *introduction* section – this section - where the challenges faced due to current planning and implementation approaches to climate adaptation are introduced; (ii) a *background* section where the theory of single, dual and multiple perspectives are explained in addition to the implementation practices and related implementation issues; (iii) the *methodology* explaining how the climate adaptation challenges from a context of flood risk management can be better analysed, together with the procedure for creating the framework ; (iv) the *framework* section, where the findings from the analysis of the literature reviewed and examples are presented in the form of a framework that can be used to help structure adaptation challenges from a multiple perspective; (v) a demonstration of the framework of flood management using a case study in Can Tho and finally; (vi) *discussions* and *conclusions* where the findings are synthesised and considered for application in other contexts and situations.

2.2 Background

2.2.1 Contemporary adaptation planning perspectives

Climate adaptation is frequently being examined through various singular perspectives – i.e based on domain specific points of view or a particular attitude or way of looking at something - such as societal, engineering, planning, economics or vulnerability reduction perspectives (Bowen et al. 2012; Kind 2014; Klijn et al. 2015; Kreibich et al. 2015; Stern 2007). Such singularity of perspective may be an example of Kant's 'Anschauung' (Carus 1892) where the means of human acquisition or receptivity of knowledge is via what is presumed to be intuition, 'normal' thinking. However, climate change does not conform with typical decision makers 'normal' problem, i.e. a problem that is recognised as fitting with a decision makers' usual intuitive view of the world.

Notwithstanding, there are also examples of studies taking a dual perspective such as those considering socio with other domains such as technical, economic, cultural or equity-inclusiveness (Birkmann et al. 2012; Kreibich et al. 2015; Newman et al. 2011; van der Brugge and Roosjen 2015). Eriksen et al. (2015) argue that there are multiple perspectives to climate adaptation and present the "Fifth Assessment Report of the Intergovernmental Panel on Climate Change" IPCC (2014d) as evidence. As illustrated in Chapters 8, 14 and 15, the report clearly recognises the need for social and political perspectives, as well as technical perspectives for effective adaptation to climate change. Such consideration of multiple perspectives requires careful attention to the: (i) progression of drivers – e.g. sea level rise, rainfall, GDP, urbanisation - that constitute the context in which adaptation happens; (ii) uncertainties attached to the projection of the drivers in the future scenarios; (iii) robustness of adaptation measures in the future across scenarios (Maier et al. 2016).

2.2.2 Challenges in structuring adaptation responses

Adaptation is complicated in ways that are not easily explored in quantitative and macro-scale studies, where the complexity of adaptation is often trivialised in the pursuit of quantification (Toole et al. 2015). The simplification of adaptation problems and responses can be attributed to: (i) the practice of analysing adaptation through singular perspectives in silos where there is a trade-off between the desire to make correct decisions and to minimize the effort (Jonas et al. 2008); (ii) Einstellung or heuristic effects - cognitive mechanisms that prevent decision makers from spending time and effort looking for alternative solutions to a problem when they believe they already have 'an adequate' one based on what has 'always been done'. This effect inhibits even experts in expending effort in examining the crucial features of the problem even when detailed information is available (Bilalić et al. 2008). Over a period of time, simplification of a type of problem in order to readily understand or grasp it, often leads to heuristic based decision-making that can embed systemic errors as well as trivialise the issues. Adaptation planning needs to be considered as a complex problem, as complexity or a wicked problem and approached using a framework such as that for a complex adaptive system (Dewulf 2013; Dunn et al. 2016; Geldof 1995; Holland 1992). This requires a systems approach, seeing "systems as a whole"; where problem structuring and solving requires an understanding not just of the components of the system, but also of their interrelationships and their relation to the whole (e.g. Young and Hall (2015), Von Bertalanffy (1972)). Approaching systems as a whole and an understanding of whole system behaviour, for example of flood risk management systems, are recognised as important steps in strategic flood risk management system understanding (Sayers et al. 2015).

2.2.3 Gap between adaptation planning and implementation

Climate adaptation in domains such as flood risk management is a conundrum, as: (i) the main adaptation interventions are long-lived, capital intensive and largely irreversible (Gersonius et al. 2012b); (ii) the decision making for adaptation is beset with uncertainties, which necessitates an approach that is flexible and in itself adaptive to the system changes and also to other changes (Anvarifara et al. 2016). Flexibility is seen as a desirable feature that enhances system capabilities and functionality (Schulz et al. 2000) and lessens the effects of maladaptation throughout the entire life cycle (Gersonius et al. 2013). In one of the largest scale attempts at flexibility in adaptation, the Delta program in the Netherlands is based on adaptive delta management. This recommends a flexible approach as a means for implementing measures in the immediate term or somewhere in the future – i.e., to speed up or defer implementation of adaptation measures, or implement other measures that can prevent the risk of over or under investment (Deltacommissaris 2014; Zevenbergen et al. 2015a). The ability to modify investment decisions is referred to as 'managerial flexibility' (Triantis 2003). Incorporation of flexibility with respect to implementation of climate adaptation measures is provided in various ways: such as allowing midterm adjustments and modifications of structure (van Buuren et al. 2013; Woodward et al. 2014); keeping investment or implementation measures open for future adaptation (Haasnoot et al. 2012b; Zhang and Babovic 2012); postponing adaptation until the time when the cost of further delay would be more than the benefits (Felgenhauer and Webster 2013).

The following barriers to flexible arrangements have been identified (Ellen et al. 2014): (i) policy makers' preference for robust arrangements; (ii) detailed contracts as modus operandi; (iii) existence of one powerful actor; (iv) divisions between policy making and policy implementation; and (v) underestimation of required implementation space. It follows from the above that interim adjustments in the implementation process may affect adaptation strategies being implemented simultaneously in other domains or spatial levels and that an *ex ante* analysis of the (potential) interdependencies between these strategies is not yet common (EEA 2016). The implementation issues continue to remain capricious, which requires understanding of the capabilities of actors; and the influence of factors such as a natural disaster or shift in markets on the actors involved (Phi et al. 2015). The lack of clear roles and responsibilities for actors in an integrated flood risk management approach is also often a barrier for the implementation of flexible arrangements (Sayers et al. 2015).

Hence it is clear that there is a gap between the adaptation planning and implementation of adaptation measures, which is a problem that impedes effective climate adaptation. These gaps are due to: (i) singular perspectives on adaptation (Vink et al. 2014); (ii) a lack of flexible implementation arrangements (Ellen et al. 2014); (iii) the trivialisation of complexities (Toole et al. 2015); and (iv) an absence of proactive analysis of implementation issues (Phi et al. 2015). The gap between policy making and implementation as well as the underestimation of implementation space can be overcome by properly structuring the adaptation planning problem, where also the measures that are selected for implementation are understood in relation to each other and to the system as a whole. Problem framing - by conceptualising the problem based on multiple perspectives - enables better understanding of the adaptation policies, gives them meaning, renders them manageable and helps in the choice of polices that are implementable (Peters 2005; Ward et al. 2004). There are a number of adaptation problem framings such as adaptation to climate change, adaptation to climate variability and vulnerability centred adaptation (Dupuis and Knoepfel 2013; IPCC 2014d). Although there could be a common understanding at a national level about the suitability of a particular climate adaptation framing, differences frequently persist at regional and local levels among the various actors in implementing consequent adaptation policies (Buuren et al. 2013; Dupuis and Knoepfel 2013; Ellen et al. 2014).

2.2.4 Understanding adaptation in a local context

Although adaptation is mainly driven by global, regional and federal initiatives (e.g. IPCC (IPCC 2014d), EEA (EEA 2016), Infrastructure Victoria (Victoria 2016a)), adaptation is mainly about the quality of local knowledge, local capacity and willingness to act at household level and local government level (Satterthwaite 2007). In adapting to floods, many households resort to autonomous adaptation practices in the short term such as moving valuables to higher levels during flooding and in the longer term, in refurbishing houses and subscribing to insurance policies to minimise losses (e.g Rozer et. al (Rözer et al. 2016)). Whereas local governments along with federal agencies resort to policy driven initiatives such as emergency responses in the short term and may invest in major infrastructure such as drainage systems and sea walls for the longer term (Stern 2007). Urban planners, sociologists and economists describe cities as self-organising systems where there is an emergent bottom-up process creating distinct neighbourhoods and unplanned demographic, socio-economic and physical clustering (Alberti et al. 2003). The presence of a local adaptation capacity and its evolution may be attributed in part to this emergent nature of neighbourhoods (Satterthwaite 2007).

Emergent neighbourhoods are the outcome of myriad interactions and emerge bottom-up due to the interaction of individual choices and actions of many human agents (e.g., households, business, governments) and bio-physical agents (e.g., climate, natural disturbances)(Alberti et al. 2003). Resilience to flooding is therefore recognised as an emergent property of individual, community or organisations in strategic flood management (Sayers et al. 2015). The individual`s or household`s vulnerability and capacity to adapt are influenced by social, economic, political and environmental factors. The complex relationship between these factors at a local level needs to be understood while analysing the particular situation in any urban area before taking up adaptation (De Sherbinin et al. 2007).

The type of adaptation measures selected depend upon how the adaptation problem is framed using framings such as adapting to climate change or adapting to climate variability or vulnerability centred adaptation (Dupuis and Knoepfel 2013). For example, adapting to the changing climate might lead to future risk reduction measures such as an increase in dike heights (SCE 2013; Woodward et al. 2014), whereas, adaptation to climate variation might lead to risk recovery measures such as insurance and post disaster relief assistance (Bek et al. 2013) or a vulnerability centred adaptation might encourage enabling low income groups to cope with floods (Revi et al. 2014), (e.g., resettlement of vulnerable populations (Quan et al. 2014)). Although these measures originate from different perspectives, there is the possibility of synergies between the adaptation measures originating from the various multi-sectoral perspectives adopted in an urban environment (Serrao-Neumann et al. 2015). Adaptive capacity and flexibility are likely to increase due to the synergy between the adaptation measures (Radhakrishnan et al. 2017b). For example, many cities in developed countries across the globe such as Bangkok and Rotterdam are now exploring the opportunity to gradually adapt the urban fabric to flooding using on-going urban renewal activity (Nilubon et al. 2016). In order to understand these synergies, the relationships between adaptation measures and drivers that act upon them have to be understood. In this, localised investments and efforts made by households and communities towards adaptation are rarely considered at a city or national level while planning for major adaptation measures (Satterthwaite 2007). There also needs to be an understanding of the interrelationships between the adaptation measures. Analysed together, the autonomous adaptation measures at household level and policy driven adaptation at city level, can reveal the (often hidden) mechanisms available for scaling up or ramping down of adaptation measures, delaying or speeding up the time of implementation of adaptation measures (i.e. flexibility). Also the consideration of the widest set of actions that reduce the probability and consequences of flooding is an important principle of strategic flood risk management (Sayers et al. 2015). This should be better explored for inherent flexibility – to implement – in an emergent urban context through pooling and analysing the adaptation measures from various perspectives and levels.

2.3 Methodology

An analysis of the current published adaptation plans through review of the literature (e.g., (Bowen et al. 2012; Gersonius et al. 2013; Kind 2014; Klijn et al. 2015; Kreibich et al. 2015; Revi et al. 2014; Stern 2007; Vink et al. 2014; Woodward et al. 2014)) provided the insights and requirements to define the individual steps in structuring an adaptation problem. Based on these findings a framework has been developed to enable the structuring of climate adaptation challenges though taking a multiple perspective. The adaptation problem structuring framework (Figure 2-1) is based on: (i) a systems approach – the systems approach has been partially used where the urban area is considered as a system, i.e., "system as a whole" in order to understand the interactions and relationships (Von Bertalanffy 1972; Young and Hall 2015); (ii) existing climate adaptation framings that look at impacts, adaptation and vulnerabilities (IPCC 2014d; UNEP 2014); (iii) contemporary approaches that are prevalent in identifying and incorporating flexibility to adapt (Gersonius et al. 2013; Haasnoot et al. 2012b; Radhakrishnan et al. 2017b; Woodward et al. 2014); (iv) proactive analysis of implementation issues during adaptation (Phi et al. 2015); and (v) inclusion of local context and integration with other planning processes (Sayers et al. 2015). To demonstrate the applicability of the framework a case study has been used: Can Tho, Vietnam. The multifarious ongoing climate adaptation plans for Can Tho in Vietnam provide an example to demonstrate the applicability of the framework. It is important to note that the focus here is on the compatibilities and synergies; and, not all the aspects of the framework are illustrated in this case study, such as those dealing with maladaptation and trade-offs.

2.4 Framework for structuring climate adaptation responses using multiple perspectives

In order to structure the climate adaptation 'response', it is essential to understand the context in which adaptation is happening - i.e. the stressors in the context, perspectives, drivers in the context, adaptation measures in these contexts and the relationship between them. The sensitivity of the system to the stressors and the link between adaptation measures can be identified from existing literature. However, exhaustive literature may not be available in all contexts. In such instances stakeholder consultations should be conducted to determine the multiple perspectives, stressors, and sensitivity modelling could be used to give a sense of what the key drivers are in the context.

A six step generic framework has been defined for context specific structuring of a climate adaptation problem (Figure 2-1) using multiple perspectives. This is further explained in the following section. The framework is set out to as a representation of a strategy where adaptation context, drivers of change and multiple perspectives help in determining the features of adaptation measures and the links between them. Although the objective of solving the adaptation problem tends to shift the focus and importance to the implementation of measures, the framework emphasises the background elements that are required for successful problem solving. The three interlocked gears - *'ascertain the adaptation context and needs, 'bring together the multiple perspectives in the adaptation context'* and *'determine the drivers of change'*, form the core of the framework; which then guides the decision maker to *'collate the characteristic features of adaptation measures',* and *'establish the links and compatibility between the adaptation measures across perspectives';* and finally culminates to' *finalise and implement adaptation measures.* The steps are numbered for sake of understanding and do not necessarily indicate the sequence. The framework can begin with step 1 and end in step 6 but alternative entry points can be selected (general flow direction is indicated by dark arrows). The framework also comprise a feedback or learning process (indicated by the white dashed arrow) from the implementation stage to the subsequent adaptation responses that could helping in reducing the gap between planning and implementation of adaptation measures.

2.4.1 Ascertain the adaptation context and needs (Step -1)

What is the context in which adaptation is taking place? Adaptation towards a single predominant stressor? Or adaptation for multiple stressors? Stressors are the set of drivers that expose the vulnerability of a community (IPCC 2013). Some of the stressors leading to adaptation are climate change, climate variability, land-use change, degradation of ecosystems, poverty and inequality (Burkett et al. 2014). As adaptation is context specific it is pertinent to find out the changes to which adaptation occurs, i.e. if the adaptation is to cope with climate change or is there adaptation to cope with other aspects as well, such as adapting to socio-economic changes and adaptation to political changes? For example, it is common to see adaptation in countries such as the UK and Netherlands to single stressors such as climate change, due to drivers such as sea level rise and rainfall increases (Gersonius et al. 2013; Woodward et al. 2014). Although definitive evidence is lacking and it is not possible to generalise too much, it may be that in typical developed countries, the socio-economic regimes are more well-developed and stable than for many developing countries. However, the relative stationarity of socio-economic stressors or the magnitude of impact from these stressors is likely to be less than that of stressors like climate change which is all pervasive. This may explain why there is a narrower focus on a single stressor, like climate change in many developed countries. In contrast, in developing countries, they are trying to cope not just with climate change but with all other aspects of a growing economy, which implicitly requires a more multi-perspective approach (Pathirana et al. 2017; UNEP 2014).

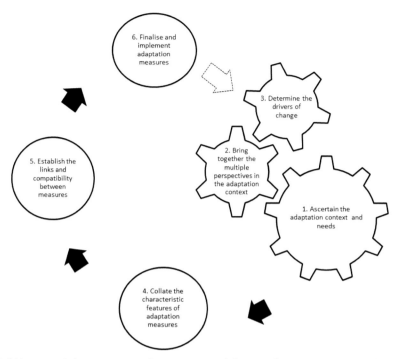

Figure 2-1 Framework for context specific structuring of climate adaptation responses. The illustration comprises a framework based on multiple perspectives as well singular perspectives. The individual steps (1-6) connected by the dark arrows constitute the comprehensive multiple perspective framework. The white arrow enclosed in black dotted lines represent the feedback and learning from the implementation stage to the subsequent adaptation responses that would reduce the gap between planning and adaptation.

2.4.2 Bring together the multiple perspectives in adaptation context (Step 2)

There may be a requirement to address multiple perspectives for adaptation even due to a single stressor. Climate risks, impacts and adaptation are location and context specific (UNEP 2014; Wolf 2011) and all of the prevailing perspectives on adaptation available in the set context – single and multiple – should be ascertained. For example if the adaptation need is driven by climate change there could be an infrastructure oriented perspective, planning perspective, socio-economic perspective, ecological perspective, political perspective, etc. Each and all of these perspectives have to be considered. The review of publications, including adaptation plans and literature would be sufficient to ascertain what the multiple perspectives were. However, this may not be adequate for all cities, especially where there may not be existing plans. For these, the identification of single and multiple perspectives could be undertaken through an exhaustive stakeholder consultation to identify as many perspectives as possible.

2.4.3 Determine the drivers of change (Step 3)

There could be a single driver (such as rainfall increase) driving adaptation or multiple drivers (such as rainfall, sea level rise, increase in population) driving the need to adapt. It is necessary to determine all the drivers that drive the change in the set context and check if the drivers are independent of each other or linked. How are these drivers affecting the urban area? How are these drivers changing over time? Is there a certainty or uncertainty about the progression of drivers over time? For example, if the adaptation in a city is towards coping with climate change impacts, the perspective could be driven by drivers such as rainfall and sea level rise that may determine the magnitude of flood depth. The same city could also adapt in parallel to economic change due to a driver such as GDP growth. The change in GDP might affect the income levels, which in turn might have a bearing on vulnerability to floods. Hence the convergence of rainfall, sea level rise and GDP growth as driver – climate and socio-economic change – need to be considered in terms of the effect on vulnerability to flooding. The effect of drivers such as river water level and urbanisation on vulnerability of households to flooding could be seen during the analysis of adaptation in Ho Chi Minh City, Vietnam (Storch and Downes 2011).

Similarly in developing countries where there is urban upgrading or socio-economic progress, there is typically a gradual improvement in housing conditions and access to social services such as healthcare or education (Garschagen and Romero-Lankao 2013) and the effect of this on vulnerability to flooding should be taken into account by considering the consequent reduction in vulnerability. The unaccounted for vulnerability reduction or enhancement can be attributed to the silo or compartmentalised thinking arising out of single perspectives. This is also true for other institutions or departments even in the same organisation, such as roads vs drainage; sanitary vs storm water; ecology vs infrastructure, etc. For example in the UK, early improvements, such as the Clean Air Acts in the 1950-1960s controlled the worst of the air pollution arising from industrial emissions – known as 'smog', whereas the less obvious but as potentially damaging to residents health that has grown in impact since then, is pollution from vehicle emissions that has not been tackled (Samoli et al. 2016) as a result of silo thinking.

2.4.4 Collate the characteristic features of adaptation measures (Step 4)

It is necessary to collate each of the adaptation measures recommended or implemented from all of the various perspectives. Determine the characteristic features of adaptation measures such as: (i) nature of measure – hazard reduction or vulnerability reduction; (ii) explicit external factors that trigger or determine the implementation and magnitude of measures; (iii) level at which a measure is being applied such as city level, neighbourhood level or property level; (iv) emergent or autonomous measures, that are not planned and supported by city or government agencies but are bottom up responses that emerge from a local adaptation context For example the dikes built to prevent the flooding of urban areas typically fall under the category of being a hazard reduction measure (one funders' responsibility). This is also a measure which is planned at a city scale and is directly influenced by the external drivers such as rainfall and sea level rise. Examples of autonomous measures emerging from a local context are temporary flood barriers around properties, elevation of floor levels of buildings in response to flooding. The conducive or constraining environment for emergence or autonomous measures depends upon the prevailing building, planning and land use regulations. At the least there are issues with the autonomy of the population and the regulatory system with regard to emergence or autonomous measures in developing countries. In developed countries this autonomy is severely constrained. For example, in Brisbane prior to the 2011 floods, a resident who was an engineer with a PhD wished to raise the level of her traditional 'Queenslander' property. She raised it the maximum allowed by the City ordinances, although she wanted to raise it further. She was flooded in 2011. After the event the City changed the ordinance to allow higher floor levels, but she had already invested heavily and could not afford to raise the property any higher (Ashley 2012).

2.4.5 Establish the links and compatibility between the adaptation measures across perspectives (Step 5)

Measures for risk reduction and adaptation require consideration of the dynamics of vulnerability, exposure and their linkages with socio-economic processes (UNEP 2014). Upon identifying the nature of each measure, the linkages between the adaptation measures across perspectives should be established in line with system approaches, such as Von Bertalanffy (1972), which recommends ascertaining the nature of the links between the various components of the systems and to the system as a whole. These linkages would also help in establishing or re-establishing the functionality of a measure through a combined perspective. Also, upon determining the nature of the measure it is possible to establish if the adaptation measure is being driven by a single driver or if there are additional or secondary drivers that might have a strong or weak influence over the measure. For example as discussed in Step 3 above, the economic driver such as an increase in GDP might lead to urban upgrading and poverty reduction, which in turn can result in the autonomous reduction of vulnerability. Further, green urban drainage infrastructure measures comprising rainwater tanks, green roofs, rain gardens, wetlands – known as Low Impact Development systems (LID), Water Sensitive Urban Design (WSUD), Sustainable Drainage Systems (SuDS) and Best Management Practices (BMPs) (Fletcher et al. 2015)– are increasingly gaining acceptance across multiple perspectives such as engineering, liveability, sustainability and resilience (Ashley et al. 2013b) and could be expected to be utilised as responses that provide multiple types of outcome and are at the same time flexible.

Compatibility between adaptation measures: Does the synergy due to combinations of measures facilitate the achievement of the adaptation objective in an effective or efficient manner? Does the combination of adaptation measures hinder achievement of the adaptation objective? The various adaptation measures collated across perspectives may be compatible. For example, from an analysis of the adaptation measures in some of the Deltas around the world (e.g. Ganges, Mekong, Chao Phraya, Rhine, Mississippi) at city and household scales (Nilubon et al. 2016; Radhakrishnan et al. 2017b; Wesselink et al. 2016) it may be concluded that: (i) the estimated functional life span of the city scale flood protection measures could be enhanced by including the coping measures that are practiced at the household level; (ii) consideration of coping capacities could be more effective in planning adaptive flood prevention measures such as spatial planning, where more extensive inundation could be accommodated in certain parts of the city.

2.4.6 Finalise and implement adaptation measures (Step 6)

When it comes to successful adaptation planning, structuring the adaptation problem is only a part of the goal. In order to have an effective outcome the measures identified have to be sequenced and evaluated in a way that is easily understood by decision makers. Adaptation pathways and Real-in options are some of the methods that allow for flexible implementation of adaptation measures in urban flood risk management as uncertainties are revealed over time (Gersonius et al. 2013; Woodward et al. 2014). Adaptation pathways may be generated using a precedence-based data specification template - description of measures and their relations in a logical and compact manner. Evaluation framework such as XLRM (Lempert 2003) provides scope for ascertaining the performance of adaptation measures or adaptation pathways over a combination of drivers. Selection of an adaptation pathway for implementation from among these pathways can then be made using set thresholds or Net present value or likelihood of occurrence among all plausible scenarios. The following section provides a demonstration of application of the framework.

2.5 Analysing climate adaptation planning and implementation in an urban context: Can Tho, Vietnam

Can Tho is one of the fastest growing cities in the Vietnamese Mekong Delta (SCE 2013). Can Tho is located on the banks of the Bassac river (Figure 2-2) and the average elevation of the city is 1.5 m above mean sea level (SCE 2013).

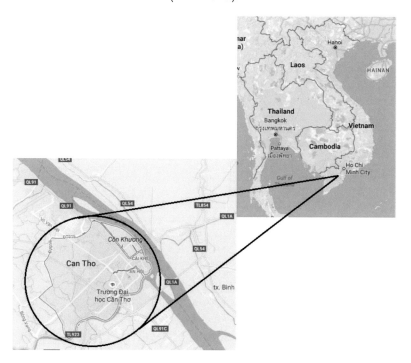

Figure 2-2 Map showing location of Can Tho (Source: Google Maps - https://goo.gl/maps/6D73uLoJYj82)

There are various singular and dual perspectives being taken and a broad body of research and other literature on climate adaptation for Can Tho. Also, there is unanimous agreement among the various studies (eg. MDP (2013), World Bank flood resilience report SCE (2013), Pathirana et al. (2017)) that there is an adaptation gap in Can Tho in responding to climate change, as the city is located in the Mekong Delta, which is especially vulnerable to climate change (IPCC 2014b).

One of the prime reasons for choosing Can Tho as the example presented here is the authors' direct experience of various issues in Can Tho such as urban flooding, drinking water supply improvements, water quality impacts and climatology issues and the formulation of the potential responses to address these. In the example, for clarity the adaptation responses considered are restricted to two groups only: the traditional institutionally led large-scale inundation risk reduction measures; and the local, typically bottom-up and emergent responses of local dwellers, raising the floor levels in their properties. These are examples of a singular 'protect' perspective and a local 'social' perspective respectively. The example demonstrates how, by taking a dual perspective from the outset and integrating both of these singular perspectives, considerably greater benefits can be obtained.

2.5.1 Ascertain the adaptation context and needs (Step -1)

Can Tho is being threatened by the consequences of climate change, as for example, a 1m rise in sea level rise would expose about 50% of the entire delta area to floods (MDP 2013). The city is also likely to be affected by an increase in river water levels due to sea level rise; an increase in river discharge due to increase in rainfall, deforestation and river training works upstream; and an increase in urban runoff due to rainfall and increased imperviousness (Huong and Pathirana 2013; Smajgl et al. 2015; Van et al. 2012; Wassmann et al. 2004). The rapid urban development in Can Tho has led to unplanned growth, increase in real estate prices, widespread water pollution and flooding issues and prevailing social disparities in terms of availability of housing stocks and access to services among the residents (Garschagen 2014). Can Tho is adapting to climate change as well as socio-economic change and to a certain extent is still adapting to the political changes after the Indo-China war (Garschagen 2014; Pathirana et al. 2017; World Bank 2014). Hence there are multiple stressors and multiple adaptation contexts in Can Tho to changing climate, urbanisation and economic situation. In countries such as Vietnam, adaptation can be seen as being driven by multiple stressors such as climate change, land use change and poverty (Garschagen 2015), as is evident from the analysis of the urban adaptation that is currently underway in Can Tho.

2.5.2 Bring together the multiple perspectives in adaptation context (Step 2)

There has been considerable climate related research for Can Tho, but there is a significant gap between knowledge and practice (Radhakrishnan 2015). Considerable research has also been conducted in the City that considers multiple perspectives – vulnerability reduction, institutional, planning, water quality, infrastructure improvements - concerning climate change and adaptation at multiple levels (Birkmann et al. 2010; Birkmann et al. 2012; Clemens et al. 2015; DWF 2011; Garschagen 2014; Huong and Pathirana 2013; Quan et al. 2014; SCE 2013; VIAP-SUIP 2013). The traditional engineering perspective is reflected in the recent flood risk management plans that are being prepared and considered for implementation, where the emphasis is more on avoiding floods by means of: (i) dike rings; (ii) improvements to drainage systems; (iii) and increasing the freeboard of flood defence systems, roads and houses by 50cm (SCE 2013; SIWRP 2011; VIAP-SUIP 2013).

In contrast with the traditional engineering perspective, the social and socio-economic aspects of flood risk management in Can Tho illustrate the experiences of living with water by the residents: tolerance to flooding; coping measures being taken at household level; direct and indirect damage to households. These are all the factors that trigger the households to implement household measures (Chinh et al. 2016a; Chinh et al. 2016b; DWF 2011; Garschagen 2014; Garschagen 2015; SCE 2013).

In spite of having a high level of preparedness: (i) the damages due to flooding in households is high in Can Tho, sometimes becoming intolerable by exceeding several months of income; and (ii) the losses due to disruption of business are higher than the loss due to the physical damage in small business establishments (Chinh et al. 2016b). This is a critical finding, as the people who are vulnerable to flooding are in the lowest income groups who cannot afford adaptation measures themselves. In this case income levels have a direct correlation with the implementation of household adaptation measures, such as elevating property floor levels (Garschagen 2014). Here we focus on the singular perspectives above, the 'traditional engineering' and the household coping measures to illustrate the need to take a multi-disciplinary perspective.

The Mekong Delta plan (MDP) (MDP 2013) looks at the climate adaptation from a Delta – regional- perspective. MDP (2013) looks at adaptation as a strategic issue where there is an interplay between climate change related water management and economic development. The plan advocates the traditional system for the delta of living with the floods that is based on controlled flooding, which is best suited for rural areas (Wesselink et al. 2016). The socio-economic aspect is not very well reflected in the adaptation measures, where the measures are mostly infrastructure and spatial planning oriented and less on the vulnerability reduction aspects.

Thus it is apparent that there are multiple adaptation perspectives – such as engineering, socio-economic, regional perspectives - in Can Tho and therefore there is the possibility to exploit the synergies between these various individual perspectives.

2.5.3 Determine the drivers of change (Step 3)

A number of external factors contribute to the flooding and hence the flood risk management of Can Tho. External factors such as sea level rise and rainfall increases lead to increases in flood depth and frequency (Huong and Pathirana 2013; Smajgl et al. 2015; Van et al. 2012). Externalities such as economic growth rates influence: (i) urbanisation which in turn has a direct bearing on increasing imperviousness and enhanced runoff in Can Tho (Huong and Pathirana 2013); (ii) economic status such as household income and number of poorer households (Jiang and O'Neill 2015; Leimbach et al. 2015). The repercussions of macro-economic growth in Can Tho may be observed in the recent trends in the population increasing due to migration, urbanisation, social disparity in terms of income and real estate prices (Garschagen 2014). Instruments, including developmental plans at a regional level, city level and community level influence the micro and macro urban attributes such as rate of urbanisation, liveability, aspirations of the people, poverty levels and capacity to adapt at multiple levels (MDP 2013; Phi et al. 2015; PM 2013; World Bank 2014). Also the aforementioned external factors are beset with uncertainty, which complicates how best to plan for overall adaptation (IPCC 2013; O'Neill et al. 2015). The list of drivers in Can Tho includes sea level rise, rainfall increase, urbanisation, household income, social disparity and developmental plans.

2.5.4 Collate the characteristic features of adaptation measures (Step 4)

A set of adaptation measures being planned and implemented in Can Tho at various levels can be elicited from the existing adaptation plans. Measures are at a city level and implemented by the State or central agencies under the umbrella of flood protection, such as increasing dike heights, improvements to drainage systems and increasing the freeboard of roads and important buildings (MDP 2013; SCE 2013; SIWRP 2011; VIAP-SUIP 2013). These measures are based on a predetermined maximum river water level in Can Tho (design value) which is based on the observed water level in the River Hau. The adaptation tipping point - i.e, the time in the future at or after which; or the value of driver at which or above, where the adaptation measures are no longer effective (Haasnoot et al. 2012b) – for these measures occur once the water level reaches the design value. A return frequency (1 in 100 Years) is built into this design value. Measures such as spatial planning and urban waterscapes are also proposed (MDP 2013). Urban development plans such as beautification of canals and lakes, resettlement of people from vulnerable areas, have been implemented in Can Tho (Garschagen 2014; Quan et al. 2014). Furthermore, a set of household adaptation measures such as elevating the floor levels, street level measures such as construction of temporary dikes during the flooding season have also been implemented (Birkmann et al. 2012). An increase in the number of households adapting to floods since 1960 has been apparent (Birkmann et al. 2012), establishing that there has been emergent behaviour resulting in autonomous household level adaptation - influenced by flood depth, income level and property ownership.

2.5.5 Establish the links and compatibility between the adaptation measures across perspectives (Step 5)

Consideration of autonomous adaptation in households of Can Tho at the planning perspective may presume this to be entirely emergent and driven by a mix of exogenous and endogenous factors. However, there are nuances in increasing the floor level of houses that are linked to physical and economic constraints. The physical constraint is the major structural modifications to doors, windows and roofs if the floors are to be raised above 50 cm, whereas the economic constraint occurs in the form of a rapid increase in costs – almost five times – between elevating up to 50 cm and above 50 cm (Garschagen 2015). There are a number of households where it is deemed essential to elevate the floor levels but they cannot do so due to their low income levels (Chinh et al. 2016b). The failure of decision makers and key agencies to understand these intricacies of physical and economic limits on adapting can lead to tensions while implementing these measures or lead to the failure to attain flood risk management objectives. Further, positive correlations have been established between the household income and elevation of floor levels of households, i.e. high income households have higher floor levels (Garschagen 2014). This insight helps in linking the socio-economic development plans such as poverty reduction (PM 2013), which aims at increasing household incomes, and this in turn could help to trigger household adaptation measures, which is a positive feedback. It is possible therefore to be more effective to directly fund poverty reduction rather than invest in major flood management infrastructure. This could be interpreted as increasing flexibility; as the decision makers have more subsequent choice by doing this to achieve the desired objectives in managing urban flooding and the measures are more widely dispersed in a number of small, local adaptations, rather than in large, irreversible and locked-in to use dikes. This dual perspective analysis reveals the inherent flexibilities that can be more effectively utilized. This is an example of how some of the adaptation measures understood by taking a multiple adaptation perspective in Can Tho can be better linked, and thus should not be considered in isolation.

In order to exploit the synergies between the adaptation measures across the perspectives, it is vital to identify the links between these measures and the external factors that act upon the links, i.e. the factors that enable, strengthen, weaken or hinder such links. The effect of an external factor such as from sea level rise on river water levels and hence the flood levels that necessitate heightening of dikes is direct, as is the effect of poverty reduction measures that reduce the vulnerability of poorer households to flooding.

Based on a traditional engineering perspective, the usefulness of a dike in Can Tho ceases when the water overtops the dike; whereas the societies' perception of flooding reveals that dwellers tolerate and cope with flood waters up to 20cm in depth within their houses (DWF 2011). This tolerance may be used to extend the usefulness of the traditional engineering measures. By for example, designing dikes to be structurally safe during overtopping. These perspectives – engineering and social- put together, can therefore increase the range of drivers against which the acceptable performance is provided (*Chapter 3*). The functional life span of the dikes can be increased as the concept of 'living with water' is embedded into the setting of performance thresholds for the larger scale infrastructure measures (Radhakrishnan et al. 2017b). This, when considered together with the recorded household floor elevation measures (Birkmann et al. 2012), postpones the tipping point of dike elevation measures (Radhakrishnan et al. 2017b). The city level dike heightening measures and the household level measures are compatible and complementary.

2.5.6 Finalise and implement adaptation measures (Step 6)

The various adaptation measures proposed for Can Tho could be considered together and assessed using adaptation pathway approaches (Haasnoot et al. 2012b). Model driven pathway approaches (e.g Haasnoot et al. (2014), Kwakkel et al. (2015)) use physically based models such as hydraulic models and scenario generation techniques to find out the most promising pathways in terms of performance robustness towards multiple objectives. Selection of an adaptation pathway for implementation from among these pathways can be made based on: (i) set thresholds such as estimated annual damages (e.g. CRIDA (In press)

); (ii) the net present value of pathways (eg. Gersonius et al. (2013)) or benefit cost analysis (eg. Aerts et al. (2014)); (iii) likelihood of occurrence among all plausible scenarios (eg. Buurman and Babovic (2016)); (iv) assessing all the aforementioned 'objectives' using multi-objective evolutionary algorithms which could avoid the narrowing down to a pathway based on aggregated objectives (eg. Kasprzyk et al. (2016)). Hence it is theoretically possible to identify an adaptation pathway that results in lower estimated annual damages and has the highest net present value for the combination of drivers that are most likely to reoccur at low intervals. This step has not been fully evaluated in this chapter and thus should be considered as a recommendation, which has to be further tried and tested in future work. The specific learning or findings in Can Tho thorough the application of the framework in the adaptation context of Can Tho is summarised in Table 2-1.

2.6 Discussion

In Can Tho, the so-far taken technical and socio-economic perspectives that are relevant to adaptation planning and problem analysis have not automatically led to multiple perspective problem structuring or identification. In response to this, the framework developed here has guided the structuring of climate adaptation problems using multiple perspectives. The multiple problem structuring framework has elucidated clear steps with which the inferences or findings from multiple perspectives could be assembled to understand the relationship between various drivers and adaptation measures. The adaptation capacities of households in Can Tho towards coping with floods would have gone unnoticed if the adaptation planning were to be done from the predominant infrastructure oriented perspective. Consideration of coping household adaptation measures from a social perspective, when combined with an engineering perspective such as heightening the dikes is likely to increase the functional life span of these. The coping capacities of households depend on their income levels and therefore cannot be presumed; however, this does give a choice to decision makers in Can Tho to invest in flood proofing of houses through subsidies or to elevate the dikes (See section 5.5).

In Can Tho adaptation is complicated when explored at a household scale due to the interplay of a number of drivers. The assumed adaptation capacities and vulnerabilities are ultimately different in ways that are not explored in many quantitative and macro-scale studies (Toole et al. 2015). It is possible that a problem structured with many perspectives would be challenging in application when using calculation-intensive methodologies like Real -in-options (Wang 2005) without major simplifications. In-spite of the increase in complexity, consideration of multiple-perspectives for adaptation is worthwhile as it contributes to clarity of understanding of the adaptation opportunities as well as acknowledging the many and various feedbacks in the system.

Table 2-1 - Multiple (dual) perspective adaptation problem strcturing in Can Tho

Problem structuring framework	Structuring adaptation problem in Can Tho
Ascertain the adaptation context and needs	• Adapting to climate change • Adapting to urbanisation • Adapting to economic change
Bring together the multiple perspectives in the adaptation context	• Engineering perspectives for flood prevention, • Social perspectives such as living with water • Overall delta management perspectives such as Mekong Delta Plan
Determine the drivers of change	• Climate drivers - rainfall increase, sea level rise • Urbanisation – change in housing conditions • Economic change – household levels, income disparities
Collate the characteristic features of adaptation measures	• Protection against floods – dike elevation and drainage improvements • Reducing vulnerability – resettlement of vulnerable population in higher areas • Coping with floods – elevating floor levels at households (autonomous and emergent)
Establish the links and compatibility between the adaptation measures across perspectives	• Household coping capacities enhance the tipping point of flood protection measures • Household autonomous measures are driven by flood levels and income levels
Finalise and implement adaptation measures	• Adaptation pathways approach to identify problems based on single or multiple objectives

The consideration of multiple perspectives can have the further benefit of revealing the 'loss of flexibility' due to negative feedbacks. Changes in socio-economic status might lead to a change in values which could affect the individual and social perception of risk, resilience and adaptive capacity (Adger et al. 2014). There are indications in Can Tho that the aspirations of people might become higher – "We want to become like Rotterdam" - and tolerance to floods in the future might reduce (Garschagen 2014). Mapping this to the appropriate socio-economic scenarios may show that as people become more affluent, their willingness to live with water in the streets may reduce (Pathirana et al. 2017). Considering this will help prevent assumptions that reliance on household level adaptation will be valid under all future scenarios. Also, this understanding will help understand any risks of maladaptation, which are less apparent when seen from singular perspectives.

The case study shows that structuring climate adaptation through multiple perspectives is possible. This frame work would be used at policy level, i.e., there is a likelihood of using the framework at a national or regional level, which considers the local adaptation context such as autonomous adaptation or emergence happening at a city or household level. However, the biggest challenge lies in operationalising the framework. The usage of the framework is likely to be effective in a common stakeholder consultation forum. When adaptation planning is driven by a planning agency or any other stakeholder who has one predominant perspective or a dominant authority there are chances for biased decisions. In such circumstances there is the possibility of normative thinking, i.e., Kant's 'Anschauung' (Carus 1892) or heuristic effects i.e., Einstellung (Bilalić et al. 2008) that would hamper the implementation of the multiple perspective framework. This is most likely in countries such as Vietnam, where the cultural practices often hinder the effective dialogue between the stakeholders belonging to various hierarchies (Mathijs van et al. 2014). This necessitates a different form of stakeholder engagement rather than a round table format, which is prevalent in western countries. Applying the framework either top-down or bottom - up might be a challenge for taking a multiple perspective approach. Although the application of the framework is very much context specific, simultaneous application of the framework at all levels could yield better results. Inspiration for operationalising the framework could be obtained from continuous stakeholder engagement processes such as learning and action alliances (LAA)(Ashley et al. 2012).

The problem structuring framework has been developed based on insights from the dual perspective analysis of adaptation in a flood risk management system and considering socio economic development and urban development as externalities. However, there are adaptation measures from other domains such as drought management and public health risk management which comprise climate adaptation that have not been considered. These domains would come under the ambit of the framework, provided the system boundary had been extended to cover the entire urban functions in the climate adaptation domain. Such an analysis would be expected to have revealed further potential adaptation measures, inter-relationships between measures, emergent behaviours and inherent flexibilities.

2.7 Conclusions

This chapter focussed on developing and demonstrating a framework for structuring the local adaptation responses using the inputs from multiple perspectives in an urban environment that is currently adapting to climate change. A framework has been created to enhance the understanding of any local adaptation context by structuring the adaptation problem through multiple perspectives. The framework differs from the normal portfolio of measures or portfolio of approaches (e.g. UK foresight(Thorne et al. 2007)) as it aims to establish the relationships between the measures across the various perspectives within the given adaptation context. The pooling together of adaptation measures derived from multiple perspectives can lead to increased flexibility by way of having a greater number of adaptation measures and increased pathways to consider. However, merely adding more adaptation measures may not automatically translate into enhanced flexibility. Enhanced flexibility is considered by means of: (i) identifying the link between adaptation measures; (ii) ascertaining the compatibility of the measures with one another; and (iii) then creating a knowledge base comprising all plausible sequences and time epochs at which the measure could be deployed based upon the unfolding of external factors. The multiple perspective adaptation problem framework can also be used to assess the risk of maladaptation and trade-offs, which is the scope for future research. The results from the case study show that multiple perspective framing of adaptation responses enhance the understanding of various aspects of adaptation measures, thereby leading to flexible implementation practices.

3 Coping capacities for improving adaptation pathways for flood protection in Can Tho, Vietnam

"வகையறிந்து தற்செய்து தற்காப்ப மாயும்
பகைவர்கண் பட்ட செருக்கு"
Verse 878 (Thiruvalluar 31 BC)

"Know though the way, then do thy part, thyself defend;
Thus shall the pride of those that hate thee have an end"
Translation based on Pope et al. (1886)

The planning and phasing of adaptation responses is essential to tackle uncertainties and ensure positive outcomes while adapting to changing circumstances. Understanding the evolution of coping and adaptation responses and their capacities is a prerequisite for preparing an effective flood management plan for the future. The aim of this paper is to determine the effect of coping capacity on longer term adaptation responses in a flood risk management system. The objectives, requirements, targets, design and performance of flood protection measures will have to be determined after taking into account, or in conjunction with, the coping capacities. A methodology has been developed and demonstrated based on an adaptation pathways approach to account for coping capacities and to assess the effect of these on flood protection measures. Application of this methodology for flood protection measures in Can Tho city in the Mekong delta shows the effect of considering coping capacity for flood protection measures and the value in delaying the occurrence of tipping points. Coping measures such as elevating property floor levels can postpone the tipping points when dikes are no longer effective. Consideration of coping capacity in the system improves adaptation responses and leads to better adaptation outcomes.

This chapter is based on the journal article *"Radhakrishnan M, Quan NH, Gersonius B, Pathirana A, Vinh KQ, Ashley MR, Zevenbergen C (2017) Coping capacities for improving adaptation pathways for flood protection in Can Tho, Vietnam Climatic Change"*

3.1 Assessment of coping capacity along adaptation pathways

The use of information about present and future changes to assess and evaluate the suitability of current and planned practices, policies and infrastructure to cope with future changes is known as Adaptation Planning (Füssel 2007). Adaptation encompasses responses to change, including climate, urbanization, demographics, and socio-economics, that are all uncertain. In order to respond and adapt appropriately to uncertainties, flood risk management systems may be considered as socio-technical systems which includes people, organisations, technical artefacts, policies, practices and procedures, protocols and framings that are deemed acceptable together with the broader contexts into which people and technical process are enmeshed (Newman et al. 2011). Coping capacities focus on the ability to maintain the system's functioning in the event of a short term, mild hazard event, in contrast to adaptation capacities that often require adjustment of system components, structures and processes in response to experienced or anticipated long term changes in for instance, frequency of occurrence of a hazard (Birkmann et al. 2010).

The widened scope of adaptation planning which is capable of encompassing the uncertainties in the period up to which the current or planned strategy is effective - i.e. when a critical point will be reached - is known as the adaptation tipping point approach (Kwadijk et al. 2010). Adaptation tipping points (ATPs) are the critical magnitude of physical drivers where acceptable physical, technical, ecological, political, societal or economic standards may be compromised (Haasnoot et al. 2012b). An ATP approach aims to assess the effectiveness of measures while planning under uncertainty, as it helps to quantify the duration for which the measures perform sufficiently well, in turn informing decision makers. As the ATP approach can consider physical, technical, ecological, economic drivers etc., the effect of these drivers and the shift in timing of ATP due to these drivers can be determined.

Adaptation pathways provide an analytical approach for exploring and sequencing a set of possible actions based on alternative drivers such as climate, land use, demographic and socio economic changes over time (Haasnoot et al. 2013; Haasnoot et al. 2011; Offermans et al. 2011). Ensuring the robustness of adaptation plans or pathways helps to overcome unforeseen circumstances in the future and to make the process of adaptation as sustainable as possible (Walker et al. 2013). Robustness as the decision criterion transforms the selection of pathways under 'deep uncertainties' i.e., where there is no knowledge or agreement about the interactions among key parameters, probability distributions of parameters (Lempert 2003). Frameworks such as dynamic adaptation pathways / policy pathways are used for robust decision making (e.g. Haasnoot et al. (2013), Kwakkel et al. (2010)). Approaches such as 'Scenario discovery' aid in planning under deep uncertainty by testing the robustness of adaptation measures which satisfy multiple objectives under various scenarios (Bryant and Lempert 2010; Groves and Lempert 2007). Vulnerability of the system or the performance of the adaptation measure (pathways) in reducing the vulnerability under various scenarios can then be assessed using statistical analysis of results obtained using simulation models (Groves and Lempert 2007; Haasnoot et al. 2014; Kwakkel et al. 2015).

The change in values and standards of a society resulting in different system performance objectives can cause a shift in timing of the occurrence of an ATP (Kwadijk et al. 2010). The ability to adapt and the boundary conditions of adaptation are not only determined by exogenous factors and availability of technology but also by societal factors such as ethics, knowledge, risk and culture (Adger et al. 2009). Within this context, flood resilience can be defined as the ability of the system to absorb or adapt to change and continue to function as expected in the face of change; where the system comprises a technical system and its interaction with human-social systems (Ashley et al. 2013a). As a consequence, human-social systems have an important role to play in adapting flood risk management systems. Hence, the ability of these systems to cope with or adapt to flooding contributes to their resilience. For example, the people of Can Tho city in the Vietnamese Mekong Delta have been living with water and accept flooding of their houses to a certain degree (up to approximately 0.20 m for about an hour) and also tolerate the restriction of movement during floods (DWF 2011; SCE 2013). This shows that the human-social system in Can Tho can cope with floods below a certain magnitude.

Although it is clear that the implementation of adaptation measures postpones the timing of ATPs, the effect of coping capacities on the adaptation pathways is unclear and needs to be understood when planning measures. Consideration of coping capacities of the system might result in different tipping points or in a shift in timing of a tipping point as human-social systems have coping capacity. There is also evidence that the capturing of feedback between short-term and long-term adaptation measures leads to the overall improvement of adaptation (e.g., Zeff et al. (2016)). Hence, the focus of this paper has been to ascertain the effect of coping capacity on the adaptation tipping points of a flood risk management system.

3.2 Methodology for assessment of coping capacity on adaptation tipping points and adaptation pathways

In order to ascertain the effect of coping capacity of a system on adaptation tipping points the following aspects must be considered: (i) the relationship between the coping capacity and adaptation measures needs to be defined; (ii) any possible change of management objectives and requirements due to the consideration of coping capacity needs to be understood; and (iii) the adaptation pathways of the system with and without consideration of coping capacities need to be determined. The effect of coping capacity on adaptation pathways can be ascertained though the comparison of occurrence of ATPs with and without the consideration of coping capacities. The methodology presented here in Figure 3-1 has been formulated based on the methodology for generation of adaptation pathways (Haasnoot et al. 2011; Offermans et al. 2011) and aims to take the effect of coping capacity into account. This methodology has been demonstrated in the context of Can Tho city, located in the Vietnamese Mekong Delta, which is adapting to climate change.

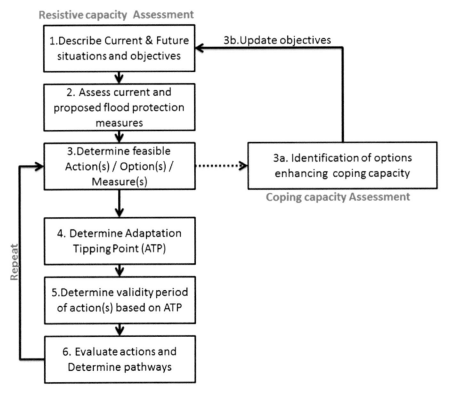

Figure 3-1 Methodology for assessment of coping capacity on ATPs and adaptation pathways

The methodology for ascertaining the effect of considering coping capacity on the adaptation tipping points for a flood system is outlined in the Steps below, with reference to Figure 3-1.

Step 1- Step 5 (Step 3a is an innovation and is explained below): Steps 1 – 5 are identical to the steps set out in the methodology for construction of adaptation pathways by (Haasnoot et al. 2011). These steps involve: 1) Description of current, future situations and objective. This includes the probable scenarios in the future that the system is expected to face, such as climate scenarios, socio-economic scenarios, land use, etc. ; 2) Analysis of the problem related to flooding such as depth of flooding, duration of flooding and effects of flooding; 3) Inventory of existing, proposed and potential flood protection measures based on the problem analysis ; 4) Determining the adaptation tipping point by modelling the effect of the external pressure such as rise in sea level, change in rainfall or runoff that are forced upon the system before and after incorporating the adaptation measure(s); 5) The validity period (time of occurrence of ATP) of a measure for a particular scenario is found by means of comparing the tipping point with the time of occurrence of the threshold when the tipping point is reached under the scenario (Haasnoot et al. 2013).

Step 3a Identification of options enhancing coping capacity: The measures taken to cope with inundation and flooding at multiple levels such as household level, street level, block level and city level should be identified by means of: (i) review of planning documents; (ii) field study; and (iii) surveys. The willingness to accept floods, the extent up to which flooding is tolerated, i.e. depth and duration of inundation; adaptation measures such as elevating floor levels of houses and embankments around districts; coping measures such as blocking of entrances during floods and staying indoors during flooding are identified in this step. The relationship between the acceptance of flooding, adaptation measures taken and the coping capacity realised through the adaptation measures need to be determined.

After establishing the relationship between the coping capacity and the measures, the impact of coping capacity on toleration of the effects such as flood depth or flood duration should be quantified. This may lead to a revision of the original objectives (Step 3b). This may also be seen as a policy revision, i.e., an adjustment due to inclusion of coping capacity and the availability of new information (Kwakkel et al. 2010; Walker et al. 2001). For example, when the floor levels of the houses are elevated by 0.20 m there is a coping capacity for a 0.20 m depth of flooding. This coping potential can be used to revise the objectives of flood protection measures, where flooding can be redefined as a situation when the depth of water is above 0.20 m on the streets, e.g. exceedance (Balmforth et al. 2006).

Step 6: The adaptation pathways and the time at which each pathway no longer provides the required service (tips) for a variety of flood protection measures can be determined by repeating the Steps 3-5. This would yield all the possible pathways and their tipping points when coping capacity is not considered. Inclusion of step 3a would yield additional adaptation pathways or enhance the effect of existing pathways where the effect of coping capacity is considered.

The effect of coping capacity on adaptation pathways is ascertained through the comparison of the ATP's of flood protection measures and their time of occurrence under various scenarios with and without consideration of coping capacities. Quantification of tipping points and the time of occurrence of individual ATPs indicates the effect and usefulness of including the coping capacity. This information can also be used to assess the usefulness of the flood protection measure under a range of drivers.

3.3 Case Study – Can Tho, Vietnam

Can Tho, the biggest and a fast growing city in the Mekong delta is located in South Western Vietnam on the right bank of the Hau river about 80 km from the sea. The Mekong Delta is one of the top five in the world that are most vulnerable to climate change, where a 1 m rise in sea level would expose about 15,000- 20,000 km^2 , i.e., about 40% to 50% of the delta region to flooding (IPCC 2007). Studies (Huong and Pathirana 2013; Van et al. 2012; Wassmann et al. 2004) show that Can Tho city will also likely to be affected by: (i) an increase in river levels due to increasing sea levels; (ii) an increase in river runoff due to climate change; (iii) an increase in urban runoff driven by imperviousness; and (iv) an increase in extreme rainfall intensity. The threats to the socio-economic development of Can Tho include seasonal flooding, sea level rise, potential land subsidence and uncontrolled rapid urbanization (World Bank 2014). Recurrence of seasonal flooding of rural areas from the Hau River during monsoons and the inundation of urban areas with water depths between 0.20 to 0.50 m on the streets are common in Can Tho. Inundation in Can Tho is due to factors such as high tide levels and the inadequate capacity of the urban drainage network during heavy rains and high river water level during monsoons. (SCE 2013). The Ninh Kieu district, which is the primary administrative and business area of Can Tho, was chosen as the study area. This area is surrounded by major waterways: the Bassac River in the East, the Can Tho River in the South and a major canal to the North. All of these water bodies are strongly influenced by tidal variations.

3.3.1 Current, future situations and objectives (Step 1)

Factors such as sea level rise, deepening of the river bed, stronger tidal action, land subsidence due to ground water exploitation, are potential causes for the increasing river levels despite there being no increase in river discharges (SCE 2013). Based on the current and future situations, the objective of the flood protection measures for Can Tho City is to ensure protection from flooding in the event of a 1% flow probability event in combination with the tides. This corresponds to the maximum river water level at +2.24m above mean sea level (Table x.9, Page 221-222 in SCE (2013)). The current master plan (SCE 2013) acknowledges that it is very difficult to completely eliminate flooding in Can Tho even with flood protection and drainage measures in place for a 1% flow probability event (page 216 in (SCE 2013). Considering this, the performance of the flood protection measures has also been defined by limiting the extent of floods within a range of 1% and 5% of total area in order to determine the adaptation pathways.

3.3.2 Problem Assessment (Step 2)

There is a lack of uniformity with respect to the design levels for dikes and embankments in the plans prepared by the various and diverse agencies responsible (SCE 2013; SIWRP 2011; VIAP-SUIP 2013), and different approaches have led to different design levels and freeboards based on 1 % flood event; year 2000 flood event; and a 1 % flood event along with high tides & 2011 rainfall in the delta. The design standards for the flood protection levels for Can Tho are based on the reference event for the flooding levels. The Prime Minister's decision No.1581/QD-TTg dated 9th October 2009, for the construction plan up to year 2015, specifies that the reference levels are for year 2000 and year 1961 flood levels in Can Tho, whereas for construction plans beyond year 2020, the reference level in Can Tho is 1 % probability flood event(SCE 2013). Further, the factors affecting flooding in Can Tho are beset with uncertainties. The rise of future sea levels and variation in rainfall and river flow will depend upon the progression of climate variables that are likely to follow one of the four plausible scenarios - RCP 2.6, RCP 4.5, RCP 6.0 and RCP 8.5 (IPCC 2013). The maximum water level during a 1% flow probability event in Hau River is likely to vary between +2.47 m to +2.83 m above mean sea level between the years 2046 and 2100 due to the sea level rise estimated by the IPCC 5th Assessment report (IPCC 2013). The uncertainty in estimates of the volume of annual precipitation for this region also varies between 8% and 29% (IPCC 2013).

3.3.3 Proposed and potential flood protection measures (Step 3)

The flood protection measures comprise alternative dikes with crests set at +2.24m (recommended by SIWRP (2011)); at +2.74m (recommended by the flood resilience master plan (SCE 2013); at +3.20m (approved by the Ministry of Agriculture and Rural Development through decision number 1721/QD-BNN-TCTL dated 20th July 2012). These are construction plans and the analyses behind these proposals do not consider the future uncertainties, hence they do not allow for flexibilities to adjust the height of the dike crest in the future. In order to account for future uncertainties an alternative, adaptive plan is considered in this paper with an initial dike crest at +2.54m that can be raised to +2.74 m and subsequently to +3.20m in the future. The following flood protection measures – dikes with flap gates at drainage outfalls – have been considered for the ATP analysis: (i) small dike – the dike with crest at +2.24m as a business as usual case;   (ii) medium dike - the dike with crest at +2.54 m; (iii) larger dike - the dike with crest at +2.74 m; (iv) very large dike - the dike with crest at +3.20 m.

3.3.4 Social acceptance and Coping with flooding (Step 3a)

Social surveys and other studies show that most people in Can Tho, including children, perceive flooding as a naturally recurring phenomenon; are aware of the risk and accept it (SCE 2013). Recent experiences have shown that there is a high level of acceptance of inundation when the houses are inundated for less than an hour with depths below 0.20 m (DWF 2011). During the 2011 floods, when inundation was between 0.20 to 0.50 m for about 2 hours twice a day, life continued normally; which illustrates that people are "living with water" in Can Tho (SCE 2013).

Individual households resort to adaptation measures that enhance their coping capacity. Measures such as the raising of floor levels are very common adaptation measures in Can Tho (Birkmann et al. 2010; DWF 2011). Many households in Can Tho build small temporary dikes in front of their houses during the floods and hence there is a higher level of acceptance of water around these houses at such times. These adaptation measures provide the necessary adaptation capacity to managing long term flooding in the streets and coping against short term flooding in the households.

There is a growing number of households elevating their floor levels. Also the upward trend of an increase in the elevation heights of building floor levels over the years is apparent (Garschagen 2014). This may be interpreted as a response to an increase of flood events over time. People are willing to raise their house floor levels by between 0.20 m and 0.50 m, but not beyond 0.50 m (Birkmann et al. 2010; Garschagen 2014). These ranges also coincide with the acceptance of flood levels and flood duration established by other studies for Can Tho (Birkmann et al. 2010; DWF 2011; Garschagen 2014). The willingness to elevate the house by only up to 0.50 m and to tolerate flooding up to 0.20 m is related to the relative cost of elevating beyond 0.50 m, which is on average some three times more expensive than the cost of elevating houses by 0.20 m (Garschagen 2014). Therefore there are two acceptable limits: a) a physical limit, that occurs when the water level in the street increases above 0.20 m and water enters into the household; b) a socio-economic limit where people are not willing to adapt or cope with the floods through this strategy due to increasing costs, which occurs when the water level in the street reaches above 0.50 m.

3.3.5 Revising objectives to reflect coping potential (Step 3b)

The planning of adaptation measures for the future and the effectiveness of current adaptation measures needed to be defined after considering the coping capacities available in Can Tho as outlined in the previous section. The original objective of flood protection, emphasizing no flooding, should be amended so as to prevent or limit the inundation of buildings by considering an acceptable limit of either 0.20 m or 0.50 m inundation in the streets. Tipping points for all the proposed and potential flood protection measures can be determined after considering the coping capacity obtained though elevating floor levels above street level by 0.20 m or 0.50 m. The 1% flow probability and floor elevations are indicators of standards and current practices that have been considered here to demonstrate the ATP approach. These standards are likely to change, as is evident from the evolution of coping capacities by households (Birkmann et al. 2010).

3.3.6 Determination of ATP (Step 4)

A hydraulic model of the study area (Quan et al. 2014) was established as the base using the SWMM5 model (Rossman 2010) based on digital elevation models, land use and existing drainage network details. The proposed and potential measures have been added to analyse the surface flooding (2D) using the PCSWMM model (James et al. 2002). The depth of inundation in the 2D cells at different water levels and for various flood protection measures were obtained from the model output which computed the areas inundated.

The adaptation measures (described in section 3.3) have been tested for flood control due to: (i) current river levels and; (ii) probable river level increases due to sea level rise based on the four future climate scenarios recommended by IPCC (2013). The effect of local precipitation has not been considered as the effect of pluvial floods is less significant compared with that of fluvial flooding in Can Tho (Apel et al. 2016). An ATP analysis showed that the current system is not effective as flooding already occurs at current sea levels.

From the ATP analysis it follows that inundation occurs at a low sea level and shifts to a higher level when coping capacities of 0.20m and 0.50m are considered. The ATP analysis for measures with a target of no flooding, limiting inundation to 1% and 5% of the study area have been carried out. The ATP for the smallest dike measure has already been reached for all of the targets considered as the current river levels are greater than +2.24 m. However, considering a 0.20 m water depth on the streets coping capacity postpones the tipping points until there is a 0.04 cm, 0.13m, 0.24 m respective increase in sea level for the targets of zero flooding, limiting flooding to 1% and 5% of the study area.

Similarly, a 0.50 m coping capacity postpones the ATPs to 0.07 m, 0.33 m, 0.42 m respective increases in sea level for the target of zero flooding, limiting flooding to 1% and 5% of the study area (Figure 3-2). The change in ATP of a measure is thus apparent when different levels of coping or alternative target limits are considered.

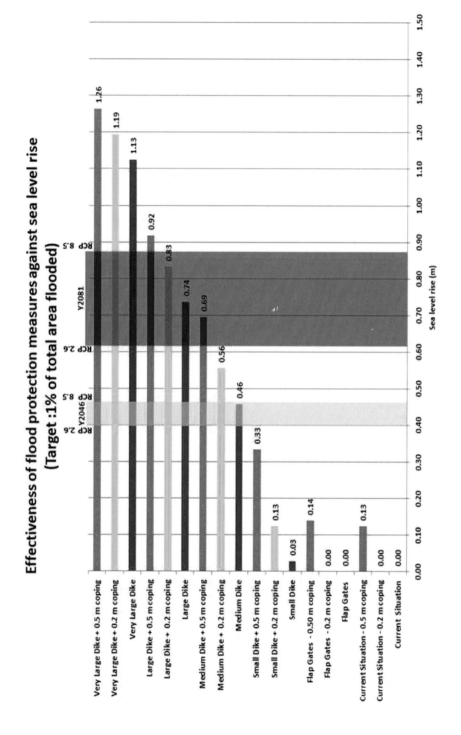

Figure 3-2 Adaptation tipping point for flood protection measures.

3.3.7 Determination of Time of occurrence of ATPs (Step 5)

The time of occurrence of ATP (validity period) of the measures can be determined by matching/ comparing the tipping point of the measure with various scenarios, i.e. the vertical bands in the light and dark shades shown in Figure 3-2. which indicate the likely range of sea level rise, from 66 % to 100% probability with *medium confidence,* i.e. based on limited evidence, but with a high level of agreement among researchers – to be expected in the years 2046 and 2081(IPCC 2013).The time of occurrence of the ATP of the small dike measure has an ATP for a 0.03 m increase in sea level, which is likely to occur around the year 2017 under all of the four climate scenarios. When the 0.50 m coping capacity is included for the small dike, the ATP shifts until there is a 0.33 m rise in sea level. The time of occurrence of ATP is likely between the years 2042 and 2048 depending on the climate scenario.

3.3.8 Creating, evaluating and sequencing Adaptation Pathways (Step 6)

A shift in the time of occurrence of an ATP could also be achieved by switching to a new measure that enhances the level of flood protection. For example in Figure 3-2, for a small dike with an ATP at 0.03 m, a higher level of protection could be achieved by switching to a 0.20 m floor elevation for all households or by implementing 0.50 m dike heightening. A later ATP could be attained by elevating the floor levels by 0.20 m and then building a medium dike or vice versa. Thus a number of Adaptation pathways may be established by combining the options and their order of implementation. Based on this concept, 91 possible pathways have been identified for the Ninh Kieu district, Can Tho, using the various flood protection measures listed in Figure 3-2. Some of these adaptation pathways for the proposed and potential flood protection measures are shown in

Figure 3-3, which was generated using the adaptation pathways generator developed by (Haasnoot and Van Deursen 2015). The coloured vertical and horizontal lines (red, blue, brown and green lines) indicate the adaptation pathways that originate from small, medium, large and very large dike measures. The transfer stations (white dots enclosed in black circles) indicate the shift from one measure to the next measure. The end of an adaptation pathway, i.e., the tipping point of a measure or a combination of measures is illustrated using a small vertical black line.

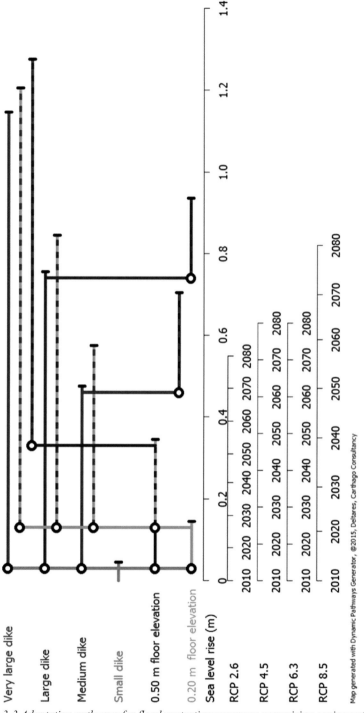

Figure 3-3 Adaptation pathways for flood protection measures comprising coping potential

From Figure 3-3, all of the individual adaptation measures and pathways have fixed tipping points with respect to sea level rise. A medium dike is effective up to a 0.46 m increase in sea level, whereas in combination with a 0.50 m floor elevation the pathway is effective up to 0.69 m increase in sea levels. The uncertainty in sea level increases are represented through four IPCC (2013) scenarios that indicate the year at which the corresponding sea level is anticipated in the future (Figure 3-3 x- axis below the sea level rise). It can be seen that the tipping points of various adaptation measures vary depending on the climate scenarios. For example, the medium dike will reach its tipping point in the Year 2070 under RCP 2.6, whereas the tipping point of the medium dike will occur in the Year 2050 under RCP 8.5. If the tipping point of the medium dike has to be extended up to Year 2070 in RCP 8.5, then a 0.50 m household floor elevation measure has to be implemented.

3.4 Discussion

People in Can Tho have been living with floods. The current master plan (SCE 2013) acknowledges the adaptation & coping measures taken by the citizens of the City. Acceptance of flooding within households and their responding by elevation of floor levels as a coping measure has been identified as a result of social surveys (Birkmann et al. 2010; DWF 2011). The inclusion of these household responses as a flood protection measure postpones the tipping point at which elevated dyke measures need to be implemented (Figure 3-3). These household measures also play a role in increasing the resilience of the system as a whole by: (i) increasing the system capacity; and (ii) reducing the impact of flooding. For example, the functional failure of the medium dike with its crest level at +2.74 m above mean sea level does not occur when the river water level reaches +2.74 m, but occurs when 1% of the area studied has a flood inundation depth above 0.20 m. Also, if elevation of household floor levels is taken up in a coordinated manner instead of the prevalent ad-hoc uptake, this would lessen the impacts of flooding in Can Tho. This highlights the importance of considering the societal aspects and social limits of the systems for enhancing adaptation (Adger et al. 2009).

Further, when short term coping interventions such as by citizens protecting themselves, are coordinated with long term adaptation interventions, like dyke heightening, this improves the overall flexibility of the system (Zeff et al. 2016). This would also be an effective no regret adaptation strategy to tackle the uncertainty due to climate change. Such no regret strategies are also in line with the recommendations of the Mekong Delta plan (MDP 2013) that advocates the need for low regret measures in order to tackle uncertainties.

Evaluation of flexible adaptation strategies are usually based on the monetary benefits (e.g., De Neufville and Scholtes (2011), Wang and De Neufville (2005)).The decisions based on a single criterion such as monetary benefits, while assessing adaptation measures or valuing flexibility could inadvertently lead to narrow outcomes. This necessitates the need for comprehensive assessment of adaptation responses. For example, including cost as one of the objective and enabling trade-offs with multiple objectives using multi-objective problem formulation (e.g. Kasprzyk et al. (2016)) is a better way of supporting decision making. Assessment of a preferred pathway or selection of a viable pathway among Can Tho's adaptation pathways (Figure 3-3) should be based on social, economic and financial costs and benefits under various plausible scenarios in the future. Also, the decision making should be based upon a multiple criteria analysis with the involvement of all the stakeholders.

It can be concluded that, in a socio-technical flood risk management system consideration of the coping capacity improves the adaptation responses and leads to better adaptation outcomes. Consideration of such coping capacities leads to a shift in time of occurrence of the ATP of adaptation measures. However, the coping capacity of a community evolves with time as a response to the adaptation measures, which is evident from the increasing heights by which floors are being raised in houses in Can Tho (Birkmann et al. 2010). Lack of proper understanding of the dynamics of the coping capacity – i.e., the increase or decrease in coping capacity with time – under the present and future circumstances may lead to a false sense of security and subsequent erosion of the coping capacity. Accounting for the evolution of the coping capacity and adaptation measures, while determining the targets for ATPs can help in realising this capacity provided by adaptation measures in the future (Garschagen 2015).

3.5 Conclusions

A methodology based on adaptation pathways has been developed and demonstrated using the flood protection measures in Can Tho. Application of the methodology in Can Tho shows that the current planning process does not consider the household coping capacity and does not take future uncertainties adequately into account. Surveys and studies conducted in Can Tho reveal that people can cope with floods as long as they are within manageable limits. Hence, formal institutional led infrastructural flood protection measures should not be seen as stand-alone measures as they consist of physical components of a flood risk management system that is inherently a socio-technical system.

Application of this methodology for flood protection measures in Can Tho city in the Mekong delta shows the effect of considering coping capacity for flood protection measures and the value in delaying the occurrence of tipping points. Coping measures such as elevating property floor levels can postpone the tipping points when dikes are no longer effective. Consideration of coping capacity in the system improves adaptation responses and leads to better adaptation outcomes.

4 Context specific adaptation grammar for climate adaptation in urban areas

"திறனறிந்து சொல்லுக சொல்லை அறனும்
பொருளும் அதனினூஉங்கு இல்"

Verse 644 (Thiruvalluar 31 BC)

"Speak words adapted to various hearers` state,
No higher virtue lives, no gain more surely great"
Translation based on (Pope et al. 1886)

In the context of climate adaptation planning there are relationships between adaptation drivers and adaptation measures, and also the interrelationships between the many potential adaptation measures. These relationships makes the selection and implementation of the adaptation measures a challenging task. This challenge may be addressed by: structuring the adaptation problem using a multiple perspective adaptation framework; and applying a context specific precedence grammar logic for selecting and evaluating adaptation measures. Precedence grammar logic is a set of rule based algorithms (grammar) that are based on the relationships in a local adaptation context, i.e., among the drivers; among the adaptation measures; and between the drivers and adaptation measures. This paper demonstrates the application of a context specific precedence grammar logic in an adaptation context – Can Tho, Vietnam – where there are multiple drivers, complex interactions between the drivers, adaptation measures and multiple futures. Adaptation pathways comprising flood adaptation measures (i.e. dike heightening) for this case were generated using rule based algorithms based on the relationships between the drivers and the adaptation measures. The results show that complex adaptation issues that are structured, can be resolved using a context specific adaptation grammar approach.

This chapter is based on the journal article *"Radhakrishnan M, Islam T, Pathirana A, Ashley MR, Nguyen HQ, Gersonious B, Zevenbergen C (Under review) Context specific adaptation grammar (algorithms) for climate adaptation in urban areas Environmental Science and Modelling"*

4.1 Introduction

Urban areas which are home to more than half the world`s population and composed of complex interdependent systems are a major challenge for climate change adaptation planning (Revi et al. 2014). The challenge and complexity is due to the interactions between social, economic and environmental stressors; where all or any can exacerbate risk to individual and to the households wellbeing (Radhakrishnan et al. 2017a). - The economic capacity and ability to make comprehensive decisions in deploying adaptation measures are seen as the key factors in determining the sustainability of Deltas, where the urbanisation and economic activities are concentrated (Tessler et al. 2015). The current frameworks on risk assessment and adaptation call for accounting of all significant natural and anthropogenic drivers in adaptation related decision making (IPCC 2014; UN 2015). This can improve the long term resilience of cities against climate change. Decision making at a programme or project level is beset with uncertainties associated with the multiple drivers (Buurman and Babovic 2016). Also there are uncertainties related to system performance in the range of scenarios anticipated in the future, and uncertainty regarding the ability of any strategy to adapt to future scenarios (Maier et al. 2016). Hence it can be concluded that adaptation related decision making in urban areas should take into account: (i) the complexity of adapting urban systems to climate change; (ii) the need for the consideration of multiple drivers, especially socio-economic (e.g. population, urbanisation, gross domestic product – GDP etc.); (iii) uncertainties associated with the drivers and; (iv) approaches set out in extant enabling frameworks for carrying out risk assessment and development of adaptation plans (Dittrich et al. 2016; Maier et al. 2016; Matteo et al. 2016; Young and Hall 2015).

Expertise on climate change, socio economic drivers that increase vulnerability and impacts, integrated assessment modelling for assessing impacts and vulnerability, is becoming increasingly sophisticated (Hallegatte et al. 2011; IPCC 2013; O'Neill et al. 2015). However, at the municipality level – the level which matters most for urban adaptation – there is a lack of enabling conditions and frameworks to support the timely evaluation of emerging urban adaptation measures that operate across a range of scales, timelines and how these are rooted in local contexts (Revi et al. 2014). There are recent decision supporting frameworks such as dynamic adaptation policy pathways (Haasnoot et al. 2013), real options (De Neufville and Scholtes 2011;

Woodward et al. 2014) and robust decision making under uncertainty, which was used in the planning of Thames estuary 2100 project (Sayers et al. 2012). There are also frameworks that approach adaptation from an investment perspective (Young and Hall 2015), which consider the performance of measures across multiple scenarios, scales and timelines. In addition to dealing with uncertainty, the strength of real options approaches applied in infrastructure domains is the consideration of path dependency (Gersonius et al. 2013). Path dependency is the dependability of the decisions made in the present on the decisions made in the past and/or the decisions that would be made in the future, that are always likely to affect the current decision. However, inclusion of path dependency in a multiple driver or multiple adaptation context and the inclusion of inter-relationships is lacking in adaptation pathways and real options approaches. Also, these frameworks do not address the complexity arising out of the relationships between multiple drivers and the interaction between adaptation measures at a finer scale such as at household level. Hence in order to help decision makers to choose and implement adaptation measures at the municipal scale, it is essential to develop an evaluation framework that is: (i) broad enough to accommodate the complexities arising out of multiple drivers; (ii) sufficiently detailed to model the interactions at finer scale; (iii) easy to understand and modifiable with a simple logical structure and ; (iv) context specific, i.e., represents the inter-relationships between the drivers and adaptation measures for the local adaptation context.

Recently devised adaptation frameworks can be used to address the concerns regarding the difficulties of including multiple drivers and adaptation across scales (e.g. Radhakrishnan et al. (2017a)). However, the detailed analysis required is always likely to be complex. The aim of this paper is to address this by showing how to overcome the challenge in modelling and evaluating a complex adaptation problem that has been structured by using a multiple perspective adaptation framework. The paper demonstrates the application of a context specific modelling and evaluation approach (Islam 2016) in an urban climate adaptation context, where there are multiple drivers; complex interactions between the drivers; relationships between the adaptation measures; and multiple possible futures.

The paper is structured as follows: (a) a review of relevant literature on flexible adaptation approaches and making the case for a context specific modelling framework; (b) methodology describing the context specific modelling and evaluation framework; (c) application of the framework in Can Tho city, Vietnam, which is currently adapting to floods due to multiple drivers; (d) discussion and evaluation of the results; and (e) conclusions of the findings specific to Can Tho and how the approach can be applied in other contexts.

4.2 The Need for context specific adaptation grammar

Modelling of path dependencies in a multiple driver context requires the understanding of various drivers and adaptation measures in a system that is undergoing adaptation. A majority of such systems are complex systems, where there are components such as variables, concepts, relationships and evaluation metrics (Hinkel et al. 2014; Ostrom 2009). The essential features of complex systems are non-linear feedback, strategic interactions, heterogeneity and varying time scales (Levin et al. 2012). Hence complex systems cannot be explained, described, predicted or modelled accurately (Cilliers 2001). For example, urban water systems can be considered as complex adaptive systems (Kanta and Zechman 2014). Urban water systems such as urban flood risk management are comprised of variables (such quantity of rainfall, river discharge) and are based on concepts such as satisfying basic services, water sensitive cities, sustainable development goals (e.g. UN (2015)). They also include relationships between variables based on deterministic relationships such as rainfall-runoff equations or non- deterministic relationships such as how the residents of the city react to flooding (e.g. Garschagen (2015)), as well as evaluation metrics such as service level bench marks (e.g. Min. of Urban Development (2017)). In the context of cities adaptation being anthropogenic; i.e. initiated by stakeholders where a rigid system can be made adaptive through adaptation measures initiated by stakeholders. Hence adaptive systems in an urban contexts are adaptable systems. Systems whose components can be modified, by decision makers, for adapting to changing circumstances are termed as adaptable systems (Oppermann 1994). Therefore complex adaptable systems can be defined as systems comprising variables, concepts and components that can be changed; the relationships among the variables and among the concepts can be established but cannot be fully explained, described or predicted accurately.

Understanding the relationships between the adaptation measures enables the decision makers to tailor any adaptation according to the needs and emergence of variables, especially the drivers. A common appraisal framework based on a system of systems approach, capable of comparing, combining and appraising adaptation measures across sectors is essential to achieve effective or efficient adaptation outcomes (Young and Hall 2015). In addition to the system of system approach, other approaches and perspectives can enhance the understanding of interaction between the adaptation measures and the drivers. For example, multiple perspectives for structuring climate adaptation (Radhakrishnan et al. 2017a) and integrating pathways (Zeff et al. 2016) are some of the recent approaches that can be used to ascertain the relationship between adaptation measures and drivers for enhancing the effectiveness of adaptation measures.

There are challenges in developing an overall modelling and evaluation framework for modelling path dependencies and inter-relationships in a multiple driver context. Firstly, integrating a system of systems approach into a modelling framework is a challenge, where a small system change can have a large overall systems effect (Maier et al. 2016). Secondly, developing a generic evaluation framework which considers change at a local scale is also a challenge as the number, nature and relationships between the drivers vary in every local context (Sayers et al. 2015; Tessler et al. 2015).   Therefore, fitting together the uncertainties and system performance in all possible scenarios is essential for evaluating adaptation measures in a multiple perspective (Maier et al. 2016). Scenarios relate entirely to changed environmental drivers (stressors) such as sea level, rainfall intensity, temperature (e.g. IPCC (2013) scenarios); and, changed socio economic stressors such as GDP, population growth rate, rate of urbanisation (e.g. Shared socio economic pathways O'Neill et al. (2015)).

Evaluating the collective robustness of various adaptation measures together, rather than of the individual robustness of a particular adaptation measure has been proposed as a way to better include uncertainty and to ensure robust system performance (Maier et al. 2016). Decision making methods to define and select robust measures are being recommended for determining the lowest level of trade-off between optimising returns (efficiency) and robustness (sustainability), although the generic toolkits for detailed analysis are still in development (Dittrich et al. 2016).

Young and Hall (2015) for example, have proposed such a generic method based on temporal and operational interdependence in a local context. However, this method needs to better include precise path dependencies, inter-relationships between drivers and the performance of adaptation measures in all plausible scenarios that are considered. Zeff et al. (2016) have proposed a risk-based framework based on dynamic adaption policy pathways (Haasnoot et al. 2013), to evaluate the performance of adaptation measures in all possible scenarios. This framework uses the concept of path dependencies and risk-of-failure as a trigger, which is the condition to switch to, an adaptation measure that will prevent the failure or the occurrence of tipping point (Zeff et al. 2016). However, in this approach the inter-relationships between the drivers and the relationship between the adaptation measures are not sufficiently detailed. The level of detailing in the inter-relationships between the drivers and adaptation, i.e., granularity, can be enhanced by means of using a context specific adaptation pathway generation and evaluation approach as developed by Islam (2016). This is further discussed in the next paragraph.

Islam (2016) has used a set of context specific rules – similar to the language grammar rules in English, French or any language that guide the formation of a sentence – for generating adaptation pathways. Precedent grammar – a term which is generally used in computer programming - is a set of production rules for generating a combination of words or symbols based upon the relationship between the words in a given set of words or symbols (Floyd 1963). Similarly, a set of rule based algorithms (grammar) that are based on the relationship among the drivers, adaptation measures and between the drivers and adaptation measures in a local adaptation context can be generated. Islam (2016) approach, termed: "*context specific precedence / adaptation grammar approach*" comprises: (i) context specific rules for generating adaptation pathways that are based on the relationship between the adaptation measures; (ii) scenario generators for creating scenario combinations based on the range of drivers under consideration; and (iii) trigger functions that are based on the adaptation objectives for switching between adaptation measures or generating adaptation pathways in a local context with multiple drivers. This paper demonstrates the application of this context specific precedent grammar approach for the generation and evaluation of adaptation pathways in an urban flood risk management system.

4.3 Methodology

Modelling context specific adaptation grammar approach (Islam 2016) was chosen to generate and evaluate pathways as it has the following advantages: (i) builds upon the dynamic adaptation policy pathway approach of Haasnoot et al. (2013); (ii) includes path dependencies and inter-relationships, unlike the generic method used by Young and Hall (2015); and (iii) is able to enhance granularity in considering the inter-relationships between the drivers and adaptation measures, unlike the risk based framework developed by Zeff et al. (2016). The context specific grammar methodology for modelling of modular systems under uncertainty developed by Islam (2016) comprises the following four components: (i) the exogenous scenario space SP influencing the system along with the number of possible scenario combinations S; (ii) the possible modules, i.e., the adaptation measures and their interrelations R; (iii) the trigger G, i.e., the control logic connecting SP and R and; (iv) the evaluation functions E based on R and triggers which are used for evaluating the adaptation pathways. Along with the components, the system time window T, time step δt and number of scenarios N should also be specified. By understanding the components in a complex adaptable urban flood risk management system, adaptation grammars, scenarios, trigger and evaluation functions for assessing adaptation measures can be generated. The terminologies and abbreviations used in the context specific algorithm are presented in Table 4-1.

Table 4-1 Abbreviations and terminologies used in the context specific adaptation grammar

a	Attribute associated with an adaptation measure, such as cost of the measure
c	Name or identification number of an adaptation measure or module
E	Value of an adaptation pathway based on the M value of the pathway across all scenarios S and time steps δt in time window T
G	Trigger function, i.e., a condition or a set of conditions used to assess the performance of the pathways or trigger an implementation measure based on the tipping point
H	Cartesian product of paths in W, scenarios S and their result for system state M at every time step δt
M	Value based on the system state of an adaptation pathway at every time

	step δt in planning range T for a pathway 'w_i' at scenario combination at all scenario combinations S_{eval} or S_{trig}
N	Number of scenario combinations generated
p	Pre-exclusion, i.e., conditions or an adaptation measure that should not be present along the pathway for implementing an adaptation measure c
q	Pre-requisite, i.e., conditions or an adaptation measure that need to be in place for implementing an adaptation measure c
R	Set of rules set describing the inter relationships between adaptation measures
r_n	n^{th} rule in the rule set R describing the inter relationship of an adaptation measure with other measures
S	Set of N number of Scenario combinations comprising all drivers in the context
S_{eval}	Scenario combinations comprising evaluation drivers
S_{trig}	Scenario combinations comprising trigger divers
S_i	i^{th} scenario combination in the scenario combination set S
SP	Exogenous scenario space comprising the adaptation drivers and uncertainties associated with them in the local adaptation context
T	Time window, i.e., Planning range
u_n	Climate or economic or demographic drivers such as rainfall, gross domestic product or population growth rate and the uncertainties associated with these
W	All possible adaptation pathways generated using rule set R
w_i	i^{th} adaptation pathway in the adaptation pathway set 'W' comprising one or more adaptation measure
x	Post-exclusion, i.e., conditions or an adaptation measure that cannot be implemented along the pathway when adaptation measure c is implemented
δt	Time interval at which an adaptation pathway assessed is using the trigger G
[ ]	Represents a set containing values such as characters or number
f ()	Represents a function

4.3.1 Exogenous scenario space

The total scenario space is defined as,

$$SP = [u_1, u_2 \dots u_n]$$

where the space is described as a collection of drivers u such as rainfall increase, sea level increase and GDP growth along with their associated uncertainties. The scenario space SP together with the planning horizon T, time step δt and number of scenarios N are used to generate a number of possible scenario combinations $S = [s_1, s_2, \dots \dots s_N]$, where $s_i = [u_{1_i}, u_{2_i}, \dots u_{n_i}]$. In local contexts where necessary, separate exogenous scenario combinations S_{trig} comprising trigger drivers and evaluation scenario combinations S_{Eval} comprising evaluation drivers can be generated. S_{trig} are the scenario combinations that are used to trigger or implement the adaptation measures, whereas S_{Eval} are the scenario combinations used in the final evaluation of pathways that are based on a user defined criteria

4.3.2 Adaptation modules

The system modules – i.e., adaptation measures – and their interactions are defined using the rule set $R = [r_1, r_2, \dots \dots r_n]$, where each rule r₁ to rₙ comprises 5 determinants,

$$r=f(c, q, p, x, a)$$

composed of a module id c, pre-requisites q, pre-exclusions p, post-exclusions x and attributes a. Let us consider for example, an adaptation measure such as constructing the superstructure for a 50 cm high dike as a module with module id $c=4.0$. The names of adaptation measures are deliberately assigned to decimal number, such as "4.0" and not as an integer or as character, such as "50_Dike" or "D50", because the context specific algorithm uses the name for generating temporary names based on the position of the measure in the pathway. The prerequisite q for this module is a condition where the foundation for the 50 cm dike is already in place or a smaller dike, say a 20 cm high dike is in place. The pre-exclusion p is the module / modules that should not be in place for constructing a 50 cm dike, i.e. if a 70 cm high dike is in place then a 50 cm dike need not be constructed. The post- exclusion x is the module / modules that cannot be implemented once the module c is in place. For example once the 50 cm high dike is in place, a 20 cm dike cannot be implemented. The attributes a comprise the characteristics which are specific to the 50 cm high structure such as cost of the dike, time for construction, lifetime of the dike, etc. These determinants in the rule set ensure that the irreversibility and path dependency of adaptation pathways that are encountered in reality are taken into account. Using rule set R a set of adaptation pathways W containing paths (w_0, w_1,.... w_i) can be generated. For example the path w_0 is pathway comprising a sequence of adaptation measures such as 20 cm high dike followed by a 50 cm high dike; w1 is a pathway starting with a 50 cm high dike followed by a 70 cm high dike; and w2 is a pathway starting with a 20 cm high dike followed by a 70 cm dike.

4.3.3 Adaptation trigger

The trigger G is the condition which determines the switch from one pathway to another. In other words trigger G is a condition leading to the tipping-point and hence the switch to a pathway where there is a likelihood for delay in tipping in the future. For example the trigger for implementing a dike could be a certain percentage of an area submerged for once in a 100 year river level. The objective of flood risk management can be to limit the area of submergence to less than 2% percent of the entire city area. For instance, a 50 cm high dike might be sufficient under current circumstances to fulfil the objectives, but increasing river levels might enlarge the area submerged to 4% in another 5 years and hence triggers the construction of 70 cm high dike in order to keep the area submerged to less than 2% of the total. The trigger can also be based on a risk based trade-off, i.e., based on cost and benefits.

The Cartesian product (W, S_{trig}) of all the scenarios S comprising generated scenarios s_1, s_2.... s_N and pathways comprising w_1, w_2.... w_n, is evaluated using the trigger function G.

$$G = f(w_i, s_i) \; \forall \; [w_i, s_i] \text{in} [S_{trig} \times W]$$

The trigger function G, determines the system state of adaptation modules, which is either 1 or 0, in pathways in all scenario combinations. System state of an adaptation measure is an attribute of the adaptation measures in the pathway, which indicates if these measures were already in place at the start of the planning period or if these were implemented during the planning range in order to postpone the tipping point. For example, the system state is '0' for any adaptation pathway at the beginning of the analysis. Let us consider a pathway w_i comprising a 50 cm dike which is later upgraded to a 70 cm dike. When the increasing water levels do not necessitate the switch from the 50 cm dike to a 70 cm dike under a scenario s_i the system state '0' is taken as the system state for scenario s_i. The system state of pathway w_i becomes '1' for scenario s_i where the increasing water levels trigger the switch to a 70 cm dike. This comparison is done at every time step for every pathway and the result are aggregated as a value M. M value is obtained for all pathways and scenarios. H is a list containing the results for the pathways evaluated using the trigger function in all possible exogenous scenario combinations S_{trig} and the result of the evaluation M at every time step t in the planning range T, *where T_S is the starting year and T_E is the ending year* .

$$M = f\left(G_{w_i\ s_i}\right) \ \forall \ \left[t_{T_s\ to\ T_E}, S_{1\ to\ N}\right] in\ [\,\mathrm{T} \times S_{trig}]$$
$$H = [M\,] \ \forall \ [\,w_i]\ in\ [W]$$

For example, a pathway containing a 50 cm high dike and a 70 cm high dike could be tested in time steps of 5 years for the next 100 years for four different sea level rise scenarios and evaluated as to when the switch occurs to the next path containing a 120 cm dike using a trigger (keeping the submerged area less than 2%). In this case there will be four results for the same pathway as there are four different scenarios.

4.3.4 Pathway evaluation function

Lastly, the evaluation function,

$$E = f(H) = f(f(m) \ \forall \ [w_i, s_i]\ in\ [S_{eval} \times W])$$

can be any summation and/or product function. As every pathway has a value M for every evaluation scenario combination S_{eval}, the pathway can be evaluated based on the value of M. For example if the pathway containing two incremental dike increases in elevation (such as 50 cm first, then 70 cm next) are capable of maintaining the flooded extent to less than 2% across all scenarios, then this pathway can be evaluated based on the timing of these two increments and the flood damages avoided by these measures. User defined comparison criteria like present value of total cost along the pathway (comprising implementation cost and flood damages), can be used to assess the appropriate timing of modules for all possible trigger timings in a pathway. In order to interpret or present the results of evaluation for the ease of understanding, a ranking criteria E, based on the present value of total cost can be calculated for each pathway so that the performance of the pathways can be compared across all possible scenarios.

Islam (2016) gives further details regarding definitions, descriptions about the sub components of the four main components – exogenous scenario space, adaptation modules, adaptation triggers and pathway evaluation function – comprising the context specific grammar architecture and for the logical structure behind the rules.

Here the context specific grammar outlined above has been applied for the evaluation of climate adaptation responses for Can Tho city in Vietnam where there is current formulation and implementation of adaptation planning for addressing changing urban flood risks and socio economic changes (MDP 2013; World Bank 2012).

4.4 Case Study

Can Tho, the biggest and a fast growing city in the Mekong delta is located in south western Vietnam on the right bank of the Hau river some 80 km from the sea. The rapid urban development of Can Tho has led to unplanned growth, increase in real estate prices, widespread water pollution and flooding issues. It also suffers prevailing social disparities in terms of the availability of housing stocks and access to services among the residents (Garschagen 2014). The city is likely to be affected by an increase in river water levels due to sea level rise, an increase in river discharge, urban runoff from rainfall, and effects of land-use change (Apel et al. 2016; Huong and Pathirana 2013; Smajgl et al. 2015; Van et al. 2012; Wassmann et al. 2004). Comprehensive analysis and structuring of the adaptation context in Can Tho using a multiple perspective adaptation framework is discussed in detail by Radhakrishnan et al. (2017a), which provides the overall guidance for the application of context specific grammar in Can Tho.

4.4.1 Structuring the climate adaptation context in Can Tho

Flooding is a recurrent phenomenon in the Mekong Delta and the people living there have long experiences of living with floods (Wesselink et al. 2016). The social and socio-economic aspects of flood risk and its management in Can Tho are evident from the community experiences of living with water: tolerance to flooding; coping measures being undertaken at household level; direct and indirect damage to the households that can be reduced because of the coping measures and preparedness due to ' living with water' ; and factors that trigger the households to implement household measures (Chinh et al. 2016a; Chinh et al. 2016b; DWF 2011; Garschagen 2014; Garschagen 2015; SCE 2013). There are multiple flood risk management plans being prepared and considered for implementation in Can Tho by various governmental and donor agencies who focus on avoiding floods by means of: (i) dike rings;(ii) improvements to drainage systems; (iii) increasing the freeboard (i.e., the vertical allowance added to the standard flood design levels to allow for uncertainty in flood levels caused by infilling or debris accumulation) of flood defence systems, roads and critical infrastructure above the maximum anticipated flood level (SCE 2013; SIWRP 2011; VIAP-SUIP 2013). Although Can Tho is making efforts at a city level and household level to adapt in the context of changing climate and changing socio-economic conditions, the true effectiveness of these measures is uncertain due to the uncertainties associated with the climate and socio-economic forcings.

The uncertainties in the climate related drivers such as sea level rise and changes in rainfall are evident from global and regional level studies such as (IPCC 2013); and delta level studies (Apel et al. 2016; MDP 2013; Smajgl et al. 2015). The uncertainties in socio-economic forcings such as GDP growth and urbanisation are evident from national level studies such as O'Neill et al. (2015) and delta level plans such as the Mekong delta plan (MDP 2013). Being as sure as possible about system performance across a range of scenarios, where there is an interplay between various drivers, is essential to tackle uncertainty (Maier et al. 2016).

The dike elevation measures (20 cm, 50cm and 70cm increase in dike heights) and household retrofitting such as raising the floor levels by up to 50 cm and beyond 50 cm are the adaptation measures that are most relevant to the local context in Can Tho (Birkmann et al. 2012; SCE 2013; SIWRP 2011; VIAP-SUIP 2013). The dike elevation measures and the corresponding construction cost of the adaptation measures have been obtained from the existing planning documents (SCE 2013; SIWRP 2011; VIAP-SUIP 2013); through consultations with the Can Tho city engineering department; and, from literature (e.g. Jonkman et al. (2013)). The dike elevation measures that are planned to be delivered by the city council are in response to rising river levels based on what is expected once in a hundred years (SCE 2013). The increase in water level of the Hau River has a strong correlation with the increasing sea levels under the various climate scenarios (Apel et al. 2016). Hence, the dike elevation measures can be considered as adaptation modules in the context specific adaptation grammar approach.

The household adaptation measures in Can Tho such as increasing the floor levels and the construction of temporary dikes are autonomous measures (Birkmann et al. 2012). According to Garschagen (2014) the correlation of adaptation actions at household level with flood depths as the driver is strong, whereas the correlation with household incomes is weak (Table 6.5, Garschagen (2014)). Also a spike in the number of houses adapting is noticed after a major flood event. These inferences are based on the responses from the households during a survey conducted in 2011. However, a trend could not be established between the recorded river levels reported by SCE (2013) and the increase in the number of households elevating the floor levels reported by Garschagen (2014). Hence, household measures are excluded here as an adaptation module in the pathways.

The increase in floor levels of houses are considered for evaluating the adaptation pathways. The avoided flood damages due to increase in floor levels and the cost incurred for increasing the floor levels are substantial and has to be considered while evaluating the adaptation pathways. Although there is a weak correlation in the relationship of household level measures to household incomes, the increase of household income and wealth is suggested as a means to increase adaptive capacity (Garschagen 2014). However the construction cost and flood damage costs at the households can vary depending upon the socio-economic scenarios. This is a context specific relationship. Hence to accommodate this context specific relationship the elevation of floor levels at households can be considered at the stage of evaluation of the adaptation pathways.

It is in this context of multiple futures of adaptation measures and interrelationships between the drivers, that the flood risk management systems of Can Tho are considered here as a complex adaptable system under uncertainty (Hinkel et al. 2014; Oppermann 1994; Ostrom 2009).

4.4.2 Application of context specific grammar in Can Tho

The Ninh Kieu district, situated along the Hau River, is the primary administrative and business area of Can Tho and this has been chosen as an important study area due to its economic importance.

4.4.2.1 Exogenous scenario space

Four climate scenarios have been considered for sea level rise (SLR) from IPCC (2013) based on representative concentration pathways RCP 2.6 – 8.5 (Table 4-2). The Shared socio economic pathways (SSP), which comprise five scenarios for GDP (Leimbach et al. 2015) – scenarios SSP1 through SSP5 – and nine scenarios for urbanisation (Jiang and O'Neill 2015) – scenarios fast fast (FF) to slow slow (SS) – for Vietnam are considered here as socio-economic scenarios. From (Table 4-2) it can be seen that the sea level increase is likely to vary between 0.44 m to 0.74 m by 2100.

Table 4-2 : Anticipated global Sea level Increase (m) under representative concentration pathways (Source: -IPCC (2013))

Year	RCP 2.6	RCP 4.5	RCP 6.0	RCP 8.5
2010	0.0426	0.0425	0.0421	0.0426
2020	0.0827	0.0817	0.0805	0.0838
2030	0.1267	0.1258	0.1218	0.1317
2040	0.1719	0.1743	0.1669	0.188
2050	0.2178	0.228	0.2171	0.2537
2060	0.2634	0.2848	0.2713	0.3294
2070	0.3089	0.3451	0.3327	0.4166
2080	0.3534	0.4067	0.4006	0.5132
2090	0.3977	0.4669	0.4731	0.6197
2100	0.4416	0.5284	0.5475	0.7362

The annual GDP growth rate which was 6.4 % in 2010 is expected to decrease and is likely to be between 0.21 % - 0.96 % in the year 2090 based on the SSP's (Leimbach et al. 2015). The urbanised population which was 30.4% of the total population of Vietnam in 2010 is expected to increase and is likely to be between 48.43 % and 91.37 % in the year 2100 based on the combination of fast, central and slow rates of urbanisation (Jiang and O'Neill 2015). Hence the scenario space *SP* for Can Tho considered here comprises the three uncertain drivers: sea level rise, GDP and urbanisation:

$$SP = [[SLR_{RCP\ 2.6}, SLR_{RCP\ 4.5}, SLR_{RCP6.0}, SLR_{RCP8.5}],$$

$$[GDP_{SSP1}, GDP_{SSP2}, GDP_{SSP3}, GDP_{SSP4}, GDP_{SSP5}],$$

$$[URB_{FF}, URB_{FC}, URB_{FS}, URB_{CF}, URB_{CC}, URB_{CS}, URB_{SF}, URB_{SC}, URB_{SS}]]$$

In urbanised population, i.e., URB, F pertains to fast growth, C pertains to central growth and S pertains to slow growth. The first suffix pertain to urbanisation up to year 2045 and the second suffix pertains to urbanisation after 2045, URBFF is a scenario where the urbanisation is fast throughout, whereas URBFS is a scenario where the growth is fast till year 2045 and slow after 2045.

A large number of multiple driver combinations based on the progression of drivers within the lower and upper bounds of the uncertainty range can be generated up to the year 2100 by analysis using monthly interval time steps. However, in order to reduce the computation effort the number of combinations N were limited to 100 at every time step. Small time steps such as hourly or daily time steps would increase the computational load whereas longer time steps such as yearly will exclude the monthly and seasonal variations. Hence the assessment time step δt considered was fixed as one month, so that the water levels in combinations reflect the monthly variations in water levels. The time window, i.e., the planning duration – is from year 2016 up to year 2100 since projections on sea level, GDP and urbanisation are available up to the year 2100. For the reasons explained in section 4.4.1: (i) sea-level rise has been considered as the only trigger driver due to the strong correlation between sea level rise and increase in river levels that triggers any need to increase dike height; and (ii) GDP and urbanisation have been considered as evaluation drivers in addition to sea level rise. Hence the trigger scenario space S_{trig} will comprise 100 scenario combinations comprising only sea levels, whereas the evaluation scenario space S_{eval} will comprise 100 scenario combinations comprising sea levels, GDP and urbanisation:

$$S_{trig} = [SLR_1, SLR_2, ...SLR_{100}] \ \forall \ [SLR_i, \delta t_i] \ in \ [SP, T]$$

$S_{eval} =$
$$[[SLR_1, GDP_1, URB_1], ...[SLR_{100}, GDP_{100}, URB_{100}]] \ \forall \ [SLR_i, GDP_i, URB_i \delta t_i] \ in \ [SP, T]$$

4.4.2.2 Adaptation modules

The adaptation measures (i.e. modules) for flood risk management for Can Tho and their interrelationships such as pre-requisites, pre-exclusions and post exclusions are presented in Table 4-3. It shows the rule set R comprising seven individual rules which represent the range of adaptation measures from a no adaptation measure state r_0 to a state where a 70 cm high dike is constructed r_6:

$$R=[r_0,r_1,r_2,r_3,r_4,r_5,r_6]$$

$$i.e.\,R = [r_0(0.0, , , ,),\, r_1\,(1.0,0.0\, ,\, [2.0,\, 3.0,4.0,5.0,6.0],[3.0,\, 5.0],19600)$$

$$................\, r_6(6.0,5.0, ,\, [2.0,4.0],204000)]$$

For example r_1 the rule set for foundation of a 20 cm dike comprises: (i) name or identification number c of the adaptation measure such as *1.0* ; (ii) pre-requisite $q -$ *0.0* i.e., the no adaptation measure condition which is a necessary condition for implementing *1.0*; (iii) pre-exclusions $p -$ *2.0,3.0,4.0, 5.0, 6.0* i.e., the adaptation measures that when implemented do not necessitate the implementation of *1.0* in the future; (iv) post exclusions $x -$ *3.0, 5.0* i.e., adaptation measures such as complete foundation for 50 cm and 70 cm dikes that cannot be implemented when *1.0* is already in place; and (v) attribute $a -$ 19600 million VND, which is the construction cost for a 20 cm dike for a kilometre.

By using the determinants for adaptation measures (Table 4-3) an adaptation pathway set W comprising pathways have been generated using the rule set R which complies with the context specific adaptation grammar (algorithm). Although there are only six adaptation modules, excluding the base case where there is no adaptation measures, it is possible to generate multiple combinations that satisfy the rules in rule set R. For example, to implement a 70 cm dike module (id *6.0*), the pre-requisite is the foundation module (id *5.0*). When a foundation module (id *5.0*) is in place, it is possible to implement a 20 cm dike (id *2.0*) or a 50 cm dike (id *4.0*) or a 70 cm dike (id *6.0*). Hence the pathway to 70 cm dike module (id *6.0*) can be a pathway with no smaller dikes or one smaller dike or two of the smaller dikes.

Thus the adaptation pathway set W comprising 78 possible pathways has been generated. A number of the adaptation pathways generated are shown in Figure 4-1 as examples. The next step after generating the pathways set W comprising all the pathways is to determine the timing of the switch from a lower dike to a higher dike for all possible scenario combinations S_{trig}.

Table 4-3: Determinants of various adaptation measures for Can Tho

Adaptation Measure	id (c)	Pre – requisites (q)	Pre exclusions (p)	Post-exclusions (x)	Attribute (a) Construction cost per km length of dike in Million VND
Current condition without any adaptation measure	0.0			0.0	
Foundation for 20 cm dike	1.0	0.0	2.0, 3.0, 4.0, 5.0, 6.0	3.0, 5.0	19,600
Foundation for 50 cm dike	3.0	0.0	1.0, 2.0, 4.0, 5.0, 6.0	1.0, 5.0	49,000
Foundation for 70 cm dike	5.0	0.0	1.0, 2.0, 3.0, 4.0, 6.0	1.0, 3.0	68,000
Superstructure for 20 cm elevation	2.0	1.0, 3.0, 5.0	4.0, 6.0		16,800
Superstructure for 50 cm elevation	4.0	3.0, 5.0	6.0	2.0	18,800
Superstructure for 70 cm elevation	6.0	5.0		2.0, 4.0	20,400

4.4.2.3 Adaptation trigger

The dikes can be implemented for the prevention of flooding at any point in time in the planning window for adapting to the sea level increase. In Can Tho, sea level rise, which leads to increased river levels, necessitates the heightening of dikes to prevent flooding (SCE 2013). Hence sea level rise is a 'trigger driver'. An anticipated increase of 40 cm of sea level by year 2090 – under RCP 2.6 - will trigger implementation of a 50 cm high dike, whereas a 62 cm sea level rise by year 2090 – under RCP 8.5 - will trigger the implementation of a further increase to a 70 cm high dike. Hence, the trigger 'G' for switching to other measures is a condition when the river level is above the crest level of a dike that is currently in place. The river level estimated is the 1 in a 100 year level based on the projected river level in the scenario combination, which is influenced by the sea level rise.

$$G = \begin{cases} 0 & River\ level < crest\ of\ dike \\ 1 & River\ level \geq crest\ of\ dike \end{cases}$$

The trigger condition G is checked at time step, i.e., every month. However, it is not practical to revise the decision to build a dike every month. In this case, the trigger condition *G* is checked for every month in a year, and the decision to switch to a suitable adaptation measure has been based on the highest water level anticipated within that year. All the 78 adaptation pathways in *W* were evaluated across 100 scenario combinations S_{trig} between the year 2016 and year 2100 at monthly intervals. The system state is 0 if G is *True,* i.e., if river level is less than crest level of the dike; or 1 if G is *False* i.e., if river level is not less than crest level of the dike at end the of the adaptation pathway.

M which is the value based on the system state *G* for a pathway at every monthly time step between 2016 and 2100 for all scenario combinations has been determined.

$$M = f\left(G_{w_i\ s_i}\right) \ \forall \ [t_{2016\ to\ 2100}, s_{1\ to\ 100}\ ]\ in\ [\ T \times S_{trig}]$$

The *M* for 78 pathways across 100 scenario combinations at yearly intervals between years 2016 -2100 were stored in a matrix *H*:

$$H = [M]\ \forall\ [\,w_{1\ to\ 7}\,]in\ [W]$$

The pathways which satisfy the performance requirement, i.e. no overtopping of the dike at any point in time during the planned for sea level increase, were then selected for further assessment.   The set of pathways that satisfy the 'no overtopping' performance requirement for Can Tho are presented in Figure 4-1. It can be seen that, out of the 78 pathways, there are ten possible pathways (numbered 0, 1-9) for implementing the various dike heightening options in Can Tho. There are pathways with a single measure such as a 70 cm high dike (pathway 2) and also pathways which contain multiple measures such as 20, 50, 70 cm dikes (pathways 1,5,6,7).

4.4.2.4 Evaluation of adaptation pathways

Once the pathways which satisfy the performance requirement are selected, the pathways can be ranked by user defined criteria. Such a relative ranking of pathways will help the decision makers to select a pathway for implementation. Relative ranking can be done with any criterion or a combination of criteria such as economic cost-benefits, eco-system benefits, travel time lost etc. In Can Tho, the present value of total cost along the adaptation pathway was considered in evaluating the pathways. Total cost comprised flood damages and cost of implementing the adaptation measures. The pathways were ranked based on the expected present value of total costs along a pathway (Figure 4-2). The flood damages along the ten adaptation pathways (Figure 4-1) which satisfy the performance requirement, were estimated using the modelling results from the PC-SWMM 1D-2D coupled hydraulic model (CHI 2017) of Ninh Kieu district developed by Quan et al. (2014) and Radhakrishnan et al. (2017b). The hydraulic simulations were run multiple times for all the pathways, for all the river water levels corresponding to the scenario combinations in S_{trig} and at all time steps δt. Using these model runs the flood depth in the case study area 'Ninh Kieu District' was determined.

The flood damages were estimated using the depth-damage relationships (SCE 2013). The flood damages were estimated at monthly time steps throughout the planning horizon for all 100 scenario combinations in S_{eval} as the flood damages depend up on the flood depth, number of houses / properties and the floor levels of households. The construction cost of the adaptation measures consists of cost of dikes and household floor elevations, whereas the damages cost comprises the road damages and household damages.

Despite the weak correlation of the household measures with GDP and river levels, the construction and damage cost for households were assumed to vary depending upon the socio-economic scenario which comprises the GDP growth rate and urbanisation; in order to consider the local adaptation context (discussed in detail in section 4.4.1). The rate of urbanisation determines the number of households at any given time; whereas the GDP growth rate determines the households that undertake adaptation measures, as household adaptation measures are directly linked to household income (Garschagen 2014). Depending upon the household incomes and

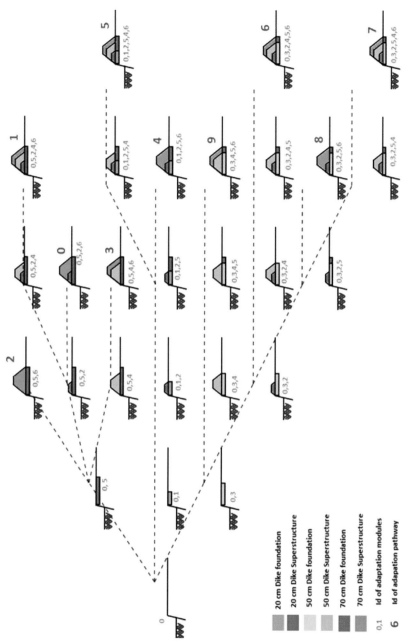

Figure 4-1- Adaptation pathways satisfying the performance criteria for all four Sea level rise scenarios in Can Tho. The adaptation pathways represented by the dashed lines comprise the adaptation measures that are implemented in sequence. The number at the end of the pathway (bold and in bigger font) indicates the name or id of the pathway. The number or numbers that are presented below each adaptation measure is the name 'c' of the adaptation measure which is also shown in Table 4-3. The implementation adaptation measure along the pathway – from left to right in the figure – is denoted by pictorially adding the measure as well as depicting the id 'c' of the measures.

exposure to flooding, many households elevate their floor levels by 0.2 m to 1.5 m (Garschagen 2014). Hence GDP growth and urbanisation were considered as 'evaluation drivers' (as opposed to 'triggering drivers').

The estimated numbers of households at any given time that: do not raise the floor levels; that raise the levels by 0.5 m; and those that raise the floor levels above 0.5 m have been used in order to compute the total construction cost and flood damages at household level. The classification of the flood adapting households into two categories – i.e. up to 0.5 m and above 0.5 m - is essential to capture the dynamics of adaptation at a household level, as any increase in floor levels above 0.5 m is five times more expensive than increasing floor levels by 0.5 m (Garschagen 2014). The pathways were evaluated based on the total flood damage and implementation cost for all the 100 evaluation scenario combinations S_{eval}.

$$E = f(H) = f(f(m) \: \forall \: [w_{1 \: to \: 10}, s_{1 \: to \: 100}] \: in \: [S_{eval} \times W])$$

The pathways are ranked in ascending order so that those with low total cost across the majority of the scenarios are the preferred path, whereas the pathways with highest total costs are the least preferred path (Figure 4-2). The pathways 0-9 are depicted along the x-axis from left to right, whereas the relative ranking of the pathways are presented on the y-axis. The least preferred pathway with lower rank – denoted by a bigger number - is at the bottom, whereas the most preferred pathway with the highest relative rank is at the top of the y-axis. The coloured circles represent the evaluation scenario combination S_{eval} and the probability of occurrence of the scenario combination. The colour of the circle represents a particular scenario combination s_i, whereas the size of the circle represents the probability of occurrence of a particular scenario. The bigger the circle, the more likely the scenario is to occur. For example from   Table 4-2 it can be seen that in the year 2080 the probability for sea level rise being less than or equal to 0.4 m is 0.75 (i.e., sea level rise is less than 0.4 m in three of the four RCP's), whereas the probability of sea level being greater than 0.4 is 0.25. Based on the values of the three drivers in SP the probabilities for occurrence for the 100 scenario combinations in S_{eval} are calculated. From Figure 4-2 it can be seen that the pathways 0-5 have higher ranking across most of the scenarios, whereas pathways 6-9 have lower ranking across most of the scenarios. Hence the decision makers in Can Tho can focus on pathways 0-5 to select the desired pathway. The other decision criteria could be the performance based on the likelihood of scenarios. Among pathways 0-5, the pathways 0, 1 and 4 have higher ranking for scenarios that are more likely to occur. Pathways 5-9 also rank low in the scenario that is most likely to occur; as the size of the dark green circles in the pathways are bigger and also ranks lower. Hence the focus can be narrowed down to the pathways 0, 1 and 4. Pathway 1 has the highest relative rank '1' in a scenario that is most likely to occur.

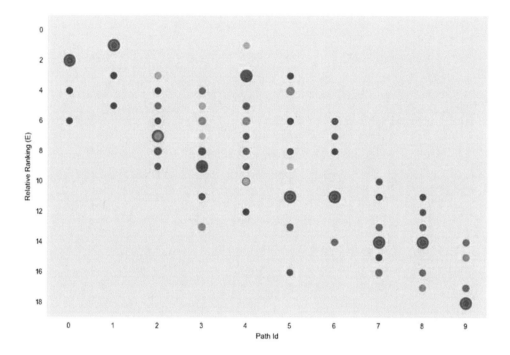

Figure 4-2- Evaluation of Adaptation pathways across scenarios based on a relative ranking comprising total construction cost and damage costs. The x axis represents the pathways 0-9, whereas the y axis represents the relative ranking with the highest rank on the top of the y axis and the lowest rank at the bottom. The position of the coloured circles along the y axis represent the rank of the pathways in a particular scenario combination s_i and the size of the circle represents the probability of occurrence of that particular scenario combination s_i. The biggest circle represents the most likely scenario combination and the smallest circle represents the least likely scenario combination.

4.5 Discussion

A performance analysis of adaptation pathways has been carried out for multiple scenario combinations using triggers such as sea level rise in order to arrive at a set of pathways that satisfy the performance criteria (Figure 4-1). The drivers such as sea level rise S_{trig} which has a strong correlation with river level increase, can be used as a trigger to implement adaptation measures and generate pathways. The other drivers which influence the local context, albeit with a weak correlation with adaptation measures, can be used to evaluate the pathways based on user defined criteria. In Can Tho, there are weak correlations between GDP growth rate and urbanisation and with household adaptation measures. Hence these drivers have been used to generate scenario combinations S_{eval} for the evaluation of pathways, which are based on the total cost incurred along the pathways. This reflects the flexibility in applying the context specific grammar as: (i) the algorithm may be modified to consider drivers with strong correlations as triggers to generate pathways and drivers with weak correlations as evaluation criteria; (ii) there is no compromise in level of detail for the relationships between drivers and adaptation measures in spite of the weak correlation.

As a strong correlation could not be established between household incomes and socio-economic drivers such as GDP growth rate and urbanisation, these were not considered as trigger drivers that determine the presence of adaptation measures along the pathways. A context specific shared socio-economic pathway (SSP) analysis of Can Tho based on specific socio-economic outcomes can be undertaken by: (i) changing the rate of poverty reduction and / or economic development in Can Tho; (ii) considering the effects of the socio economic development plan (PM 2013); (iii) changing the rate of urbanization, rate and type of industrialization recommended in Mekong Delta plan (MDP 2013); (iv) consideration of vulnerability (Birkmann et al. 2012) and climate change impacts (IPCC 2013).

The results for pathways illustrated in Figure 4-2 are based on the relative ranking derived using the expected present value of construction cost and flood damages that are likely to occur along a pathway for all possible trigger timings. The ranking of pathways when presented together with the likelihood of the scenario combinations will help the decision maker to understand the results more easily. Although not presented here, a combination of criteria such as minimal construction cost, avoided flood damages and eco-system benefits would convert the climate adaptation problem into a multiple objective problem and help in choosing pathways that satisfy all these objectives. Approaches followed in benefit evaluation tools such as BeST (Horton et al. 2016) which does not rely only on economic inputs, could be used for creating the composite relative ranking criteria. This will also resolve the Arrow's paradox of narrowing down on solutions based on a single objective where there are also other objectives to be achieved (Kasprzyk et al. 2016).

The major limitation of this methodology is that the large number of adaptation measures in a major adaptation planning process have to be neatly quantified into smaller discrete actions that can be selected at different times. Hence a careful pre-planning is necessary for applying this methodology.

Also this methodology can be extended to improve flood warning times and determine the setback distances in flood plains. However this needs further research as the inter-relationships and stakeholders involved in flood warning and in improving setback distances are different. The key towards operationalising the modelling and evaluation approaches is to account for stakeholders different beliefs regarding the levels of uncertainty, as this leads to better understanding of causal relationships and preference of management solutions (Dewulf et al. 2005). The challenge of integrating multiple perspectives, system approaches and involvement of stakeholders can be overcome by applying context specific precedence grammar logic for developing an overall modelling and evaluation approach, thereby enhancing the effectiveness of adaptation across the sectors.

4.6 Conclusions

A context specific adaptation grammar has been used to generate and analyse climate adaptation pathways to manage flooding in Can Tho. Using the methodology, adaptation pathways have been generated based on relationships between the measures that were shortlisted based on the performance of measures across scenarios, and further narrowed down based upon the criteria preferred by the decision makers or stakeholders (Table 4-3). These relationships have been used for creating a set of rules 'R' based on the precedence grammar logic to generate adaptation pathways, and to assess the collective performance of the adaptation measures (pathways) over a range of scenarios as set out by Islam (2016). Decision makers at various levels – households, city council, provincial, national and international levels – may use changes such as rising river water levels, damage to households and increasing household incomes to trigger adaptation measures (e.g., dike heightening, drainage improvements, spatial development plans and retrofitting of houses).

It has also been demonstrated that the relationship and interdependencies between the adaptation measures can be captured using determinants such as pre-requisites, pre-exclusions and post- exclusions (Table 4-3). These determinants have been used to generate pathways using rule-based algorithms, which is the core of the context specific adaptation grammar approach. An example of how the inter-dependencies can be used for generating pathways has been presented in section 4.4.2.2, for a case of building a 20 cm dike followed by a 70 cm dike.

The aim of this paper was to overcome the challenge in modelling and evaluating a complex adaptation problem that has been structured by using a multiple perspective adaptation framework. Thus this paper demonstrates the application of a context specific modelling and evaluation framework (Islam 2016) in the urban climate adaptation context, where there are multiple drivers, complex interactions between the drivers, adaptation measures and multiple futures. Application of the context specific adaptation grammar (Islam 2016) in Can Tho reveals that it is possible to generate adaptation pathways using rule based algorithms based upon the relationship between the drivers and the adaptation measures. It also shows that a context specific categorisation of drivers is possible, i.e., the various drivers used for generating scenarios can also be segregated into trigger drivers (e.g., sea level in Can Tho) and evaluation drivers (e.g., GDP growth rate and urbanisation in Can Tho). Finally, there is scope for assessing adaptation pathways based on user defined criteria, such as economic objectives or total costs along the pathways or non-economic objectives in combination with other assessment tools (e.g. BeST Horton et al. (2016)). The results from the case study show that the complex adaptation problems structured using a generic framework can be addressed using a context specific adaptation grammar approach.

5 Flexible adaptation planning for Water Sensitive Cities

"நோய்நாடி நோய்முதல் நாடி அதுதணிக்கும்
வாய்நாடி வாய்ப்பச் செயல்"
>   Verse 948 (Thiruvalluar 31 BC)

>   "Disease, its cause, what may abate the ill:
>   Let physician examine these, then use his skill"
>   Translation based on Pope et al. (1886)

Cities have started adapting to uncertain climate drivers such as temperature and sea level rise, and some cities are also transitioning towards concepts such as Water Sensitivity. In adaptation planning, flexibility is considered as an important characteristic to respond to changing circumstances. This chapter presents a novel approach to identify where flexibility can best be embedded in urban flood risk management systems. The identification of a flexible water sensitive adaptation response is based on change propagation; i.e. the response's ability to minimise negative or maximise positive *impacts* in urban *systems*. The Flexible adaptation planning process (WSCapp), comprising change propagation – especially how positive and negative impacts propagate in an urban environment, can be used by those concerned with urban planning and urban adaptation to identify *'where'* the flexible adaptation responses can be implemented. WSCapp can be used to decide the type of adaptation response such as changes to streetscape, place making or architectural forms that can best contribute towards the objectives of a water sensitive city.

This chapter is based on the journal article *"Radhakrishnan M, Ashley R, Gersonius B, Pathirana A, Zevenbergen C (Submitted) Flexible adaptation planning process for water sensitive cities Cities"*

5.1 Introduction

Adapting Adaptive approaches in planning, design and implementation can help to minimise the hazardous effects of climate change and explicitly allow for the uncertainties associated with these in urban areas (Revi et al. 2014). Policy makers, planners and others managing urban areas have recognised the likely effects of climate change and have initiated strategic adaptation actions that are aligned usually according to a particular vision used by a particular sectoral or service provision (Chu et al. 2017; Jabareen 2013). There are also signs of a breakaway from a sectoral vision, in which urban adaptation planning is compartmentalised, with moves towards multi-sector and multidisciplinary planning approaches to better bring about sustainable development (Malekpour et al. 2015). However, decision making related to adaptation faces uncertainties, which necessitate a flexible approach that can adapt to the changes. Flexibility is important for this and is here defined as the attribute of a system which enables the system to respond in an efficient way in terms of performance, cost and time, when the system is confronted with uncertainties, negative consequences and opportunities (Anvarifara et al. 2016).

Figure 5-1 4-RAP model of available strategies – where the 4R's signify the retain, relieve, resist and retreat strategies; 'A' signifies accommodate strategy; and 'P' signifies prepare strategy - to enhance flood resilience (Adapted from Gersonius et al. (2016))

Flexibility is increasingly seen as a desirable feature that enhances system capabilities and functionality in the face of uncertainty (Schulz et al. (2000). Gersonius et al. (2016) recommend flexibility in combining different types of strategies "retain, resist, relieve, retreat, accommodate and prepare" (4RAP) to increase resilience towards flooding in designing and planning systems for water sensitivity (Figure 5-1). For example, the City of Melbourne`s resilience strategy considers flexibility as an important characteristic to respond to changing circumstances when using a mix of strategies such as adapt, survive, thrive and embed (City of Melbourne 2016). Flexibility is also a property which counters the effects of maladaptation throughout the entire life cycle by allowing system change (Gersonius et al. 2013a).

A "Water Sensitive city" (WSC) vision (Brown et al. 2009) considers urban water management from a perspective of intergenerational equity and resilience to climate change and hence is more than just Water Sensitive Urban Design (Ashley et al. 2013b). The WSC approach recommends an urban design that reinforces 'water sensitive' behaviours. This is evident in the adaptation plans and actions taken by cities such as Rotterdam, Copenhagen, Dresden and Melbourne (City of Melbourne 2016; EEA 2016) and is an aspiration for London (HM Government 2016). The adaptation measures in these cities are termed as 'transformational adaptation measures' (EEA 2016). They use behaviour and technology to change the performance of urban systems fundamentally. In addition, transition or strategic planning for sustainable development requires a proactive planning culture in order to create conditions for change to deal with future issues (Malekpour et al. 2015). For example, in Melbourne, water, wastewater and storm water management was formerly aimed at the protection of waterway health, tackling water shortages during drought, ensuring water supply through alternative sources and protection against flooding (Ferguson et al. 2013a). Now Melbourne has moved beyond this and is including the objectives of being resilient to climate change and becoming a water sensitive city (City of Melbourne 2016). An important characteristic of a resilient city is flexibility, e.g, having a number of alternative ways to provide services and respond to changing circumstances as these arise (City of Melbourne 2016). It allows the city to respond to future needs from climate change as well as changes in objectives.

An effective WSC requires a process that incorporates flexibility into planning, implementation and operation. The context-first approach adaptation planning process (e.g. Thames Estuary project TE2100) makes adaptation flexible using a high level route map of adaptation measures (Reeder and Ranger 2011). Techniques such as real in options (RIO) (e.g. Woodward et al. (2014)) value the flexibility built into a (flood risk management) system in monetary terms. However, these approaches do not identify the optimal places where flexibility can be embedded. Hence, in addition to the value it is necessary to know where, how and when to incorporate flexibility to achieve the objectives of a WSC.

This paper develops a novel approach to identify where flexibility can best be embedded in urban flood risk management systems. This has been developed by drawing on knowledge and procedures from the automobile and aerospace industries, where flexible adaptation planning is everyday practice (Suh et al. 2007). The flexible physical components are selected based on the components' ability to propagate change in the urban system (Eckert et al. 2004). An adaptation response is an ideal flexibility 'candidate' when it minimises negative impacts or maximises positive impacts throughout the area under consideration (i.e. change propagates throughout the system) and not just in the vicinity of the adaptation response. For example, a dewatering pump reduces flooding in a neighbourhood (reduces negative impact), whereas green roofs in addition to reducing the peak flow during rain, also have ecosystem service benefits in the neighbourhood (increases positive impact). Both these urban water management adaptation responses are capable of propagating change throughout the neighbourhood either by reducing negative impacts or by increasing the positive impact and contribute towards increasing resilience in the urban system. Further, prior identification of flexible adaptation responses makes the response to change rapid, i.e. making the change process agile (Pathirana et al. 2017a). In this context, agility is defined as the ability of the adaptation system to respond quickly to uncertainties, threats and opportunities.

The sections in the chapter explain: (a) the relevance of flexibility in flood risk management; (b) methods that are used in embedding flexibility in the manufacturing sector; (c) the need for a planning process that ensures that adaptation is flexible in a WSC context; (d) development of a flexible adaptation planning process (WSCapp) for identifying WSC elements or components where flexibility can be embedded; and (e) theoretical and practical considerations for applying this flexible adaptation planning process.

5.2 Flexibility in contemporary flood risk management practices

Flexibility is often considered as a valuable capacity to cope with uncertainty and change, although there is no consensus about what constitutes flexibility in literature or practice (Anvarifara et al. 2016). It is context and domain specific. Flexibility applies to both the planning/design process itself as well as to the attributes of the key components or artefacts being planned and designed. For example, the Delta programme in the Netherlands is based on adaptive delta management and recommends a flexible approach as a means for creating options in terms of implementing measures in the immediate term or in the future – i.e., speeding up or deferring the implementation of adaptation measures, or implementing other measures that can prevent the risk of over or under investment (Deltacommissaris 2014; Zevenbergen et al. 2015). The incorporation of flexibility with respect to implementation of climate adaptation measures is provided in various ways as illustrated in the following examples: allowing midterm adjustments and modifications of structure – structural flexibility (van Buuren et al. 2013; Woodward et al. 2014); keeping investment or implementation options open for future adaptation – managerial flexibility (Haasnoot et al. 2012b; Zhang and Babovic 2012); postponing adaptation until the time when the cost of further delay would be more than the benefits – also managerial flexibility (Felgenhauer and Webster 2013). There are also other approaches such as functional flexibility, operational flexibility and strategic flexibility (Radhakrishnan et al. (2016).

The context-first approach adaptation planning process (e.g. Thames Estuary project TE2100; Reeder and Ranger (2011)) identifies the need for flexibility and incorporates it in the form of a high level road map of adaptation measures. Context-first adaptation approaches (e.g. Ranger et al. (2010), Dessai and Sluijs (2007)): (i) encourage the decision makers to begin at the level of the adaptation problem (or opportunity); (ii) specify the objectives and constraints; (iii) identify appropriate adaptation strategies; (iv) and only then appraise the desirability of adaptation measures against a set of climate change projections. The adaptation pathways approaches (Haasnoot et al. 2012a), dynamic adaptation policy pathways (Haasnoot et al. 2013) and model based adaptation pathway approach (Kwakkel et al. 2015) also fall into the category of context-first adaptation planning processes (Veerbeek et al. 2016). However, these context-first approaches fall short in identifying the adaptation responses of the system, especially where flexibility can be incorporated.

The real in options (RIO) techniques applied to urban flood risk management (UFRM) help to specifically respond to exogenous uncertainties and to value flexibility. For example, Gersonius et al. (2012); Gersonius et al. (2013a); Woodward et al. (2014); Zhang and Babovic (2012), focus on creating flexible alternative designs and determining the cost of alternative designs. These approaches do not identify where, when and how to embed flexibility. This may be attributed to the reductionist way of approaching complexity from a traditional engineering perspective and the consequent managerial approach (Fratini et al. 2012; Malekpour et al. 2015). The RIO application in the UFRM domain has to be customised to include the comprehensiveness of a WSC approach. Liveability and resilience as part of sustainability are the essential aspects of WSC (Ashley et al. 2013b).   Although RIO methods address resilience and sustainability to an extent (e.g., Zhang and Babovic (2012)), they do not fully address the aspects of liveability. Where liveability — the most visible and appreciated quality of any city — is the totality of features that add up to the citizens` quality of life; comprising their experience of the natural and built environment, together with services (Salama and Wiedmann 2016).

5.3 Flexibility in Manufacturing

Addressing the combination of strategy transitions and uncertainty of drivers is a challenge that is not specific only to the domain of urban water management. Similar challenges have been faced by manufacturing industries and software developers (Bernardes and Hanna 2009; Koste and Malhotra 1999; McGaughey 1999; Sánchez and Pérez 2005). Businesses and organizations such as these face a volatile environment that is highly uncertain with challenges such as increased competition, globalized markets, technology obsolescence and dynamic customer requirements. The strategy transitions in urban water management can be compared with the evolution of new car models. For example, the change of focus from reducing pollution and conserving water in urban water management to sustainable water management can be compared with the change in preference for cars with more power towards the preference for cars which pollute less or electric cars.

The automobile manufacturing sector uses product platform strategies to save costs by sharing core elements among different products in a product family (Simpson et al. 2006; Suh et al. 2007). Suh et al. (2007) have developed a seven step flexible platform design process (Figure 5-2) to deal with uncertainty and product variants in order to ensure continuity in production and maximise profit for the automotive industry. The uniqueness of this approach is the identification of the most appropriate subset of physical components, i.e., the parts of the vehicle or machine, where the necessary flexibility can be embedded (step IV in Figure 5-2). The identification of flexible adaptation responses is based on the magnitude of change that the components propagate throughout the system when modified (Eckert et al. 2004).

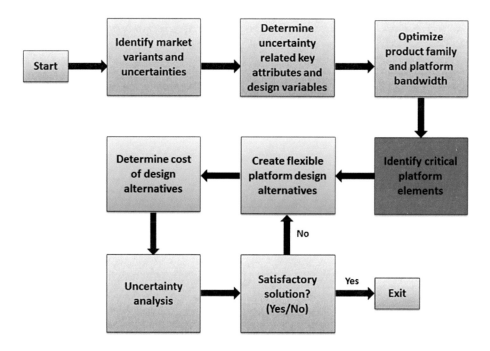

Figure 5-2: Flexible platform design process in the automobile manufacturing sector (Adapted from Eckert et al. (2004))

According to Eckert et al. (2004) the physical components that are capable of propagating greater change are to be assessed carefully before being selected as candidates for embedding flexibility. For example, in the design of an automobile chassis the width is kept constant for all the variants to limit change propagation. This is because change in width has a direct bearing on the stability of the vehicle. One of the parameters that determines the chassis dimensions - the length and breadth - is the dimensions of the engine, which in turn depends on the number of cylinders. Hence the width of the chassis is generally increased and fixed for all future variants, by taking into consideration the possible increase in width of the engine, thereby avoiding major impacts on the stability of the vehicle. However, the length of the chassis is designed to be flexible so that it can be changed in accordance to the dimensions of the new engine as the impact on stability due to change in length is minimal. Enabling flexibility in such design parameters requires initial investment in design, tooling and assembly equipment and is similar to using real in options (RIO) analysis in chassis design (Suh et al. 2007). Wherever possible, change propagation is limited by means of providing sufficient margins or excess headroom (i.e. redundancy) in the design parameter to limit the change propagation within these systems. Provision of redundancy is used when the engineering cost of design changes, additional fabrication, assembly tooling and equipment investment to make these changes are higher than the cost of designing and manufacturing a product with a significant degree of redundancy. Change propagation is similar to "Building information modelling" used by architects and planners where the effect of changes made to the physical elements such as wall or windows on lighting, ambience, spatial relationships, interaction between the various architectural elements; as well as the effect of changes in the fabrication and construction of physical elements (Lu and Korman 2010).

5.4 Flexible adaptation planning process in WSC context

In a WSC context, various urban water systems (and other urban systems) are interconnected and change in the positive or negative impacts in one system will propagate to other systems. A change rarely occurs in isolation and impacts other systems. This change propagation is important as it may have unintended consequences affecting the functionality of the systems, such as non-compliance with design standards, including flooding or increased flood damages; or a reduction in liveability aspects due to the presence of stagnant water. The components or sub systems in the WSC have to be analysed from the aspect of change propagation.

The adaptation responses that are capable of propagating more change (i.e. positive or negative impacts) when modified are the critical responses and thus potential candidates for incorporating flexibility. Change propagation can be used to assess the crucial changes (positive and negative impacts) that might propagate in the urban water system due to changes in components. For example, the use of Sustainable Drainage Systems (SuDS) can be analysed using this concept of change propagation in Australian urban catchments where there is risk of flooding and droughts. SuDS are more flexible than buried piped systems (Ashley et al. 2015). The change in the type, size or number of SuDS components - such as green roofs and rain barrels - will have positive as well as negative impacts on increasing use or size of SuDS measures: (i) reduces runoff (i.e. reduction of negative impacts throughout the catchment); (ii) can lead to downsizing or deferring the expansion of conventional downstream measures such as large-scale detention systems or pumping arrangements (i.e. increase in positive impacts in the catchment and in cities investment planning); (iii) can yield desirable benefits such as enhanced aesthetics, improvement in water quality, etc., (i.e. increase in positive impacts in the catchment and in downstream catchments); (iv) but can lead to undesirable effects such as odour nuisance or more mosquitoes (i.e. increase of negative benefits throughout the catchment); and, (v) can increase the risk of fire during droughts due to dry vegetation that necessitates additional watering and strain on water resources (i.e., propagation of negative impact even beyond the catchment).

Similarly change propagation analysis can be undertaken for conventional infrastructure measures such as the pumps at the outfall, where the reduction in negative impacts such as reduced flooding or high dependence on SuDS might propagate upstream depending upon the capacity of pumps. Also there are no other positive changes which propagate from pumps such as the additional benefits from SuDS measures during normal operations. However, change propagation can also be used from the perspective of controlling or minimising the change that propagates in the system. The limitation of a pump can be turned into an advantage as it is a standalone centralised component where the incorporation of flexibility into a pumping station, by means of additional bays for pumps for the future, is easier to implement. Also this might be economically cheaper than investments in distributed SuDS measures upstream. Thus change propagation can help in identifying the main flexible adaptation response locations (where) and measures for maintaining flexibility in future adaptation response (which). This is a balancing act between the (unwanted) physical suppression of future change propagation, where it is not wanted and investment in flexibility where this is wanted.

Consideration of change propagation due to the transition in vision – such as the transition of a city from a 'water supply city' to a WSC (Brown et al. 2009) – is as equally important as the consideration of change propagation due to any change in system components. Exploring the design, implementation and maintenance of the water sensitive or flood resilient systems such as water plazas (e.g. place making) or water retaining pavements (e.g. streetscape design) that are implemented in cities such as Copenhagen and Rotterdam (EEA 2016), can help in understanding how change – in architectural features, aesthetics, amenities, built environments, open space, streetscape – propagates in a complex urban environment. For example, such studies can help cities to understand change propagation when the required functions of, for example, the use of a wetland system changes in regard to societal needs as a transition from water quality improvement or drought resistance to a broader WSC perspective. Change propagation can be examined through mapping the relationship between the systems or the adaptation measures. For example, in order to improve the resilience of infrastructure, the State of Victoria in Australia is pooling together various infrastructure options (Victoria 2016a). How each option works with others, in terms of how they might enable, complement or inhibit one another in advancing one or more of the needs is referred to as relationship mapping (Victoria 2016b). Such relationship mapping between adaptation measures is a good starting point for ascertaining change propagation in urban water or other systems.

5.5 Development of flexible adaptation planning process for WSC

Inspiration is drawn here from the flexible platform design processes and complex change management processes (Eckert et al. 2004; Suh et al. 2007) that are prevalent in the automobile manufacturing sector in order to customise the context-first approaches to the WSC context. The adaptation planning process for incorporating flexibility in a WSC context using the principles from 'context specific adaptation approaches' and the flexible platform design process is presented in Figure 5-3. The adaptation planning process presented here resembles or is very similar to the urban adaptation planning process such as Dynamic adaptation planning process – DAAP - (Haasnoot et al. 2013) and CRIDA steps in adaptive management cycle (CRIDA In Press). In addition to the stress test and performance of the responses recommended by these adaptation planning processes (e.g. Kwakkel et al. (2016)), the emphasis in WSCapp is on changes in the resilience, liveability and sustainability aspects in a WSC context that are propagated by the responses in the urban environment (Step 4 of Figure 5-3). Hence in addition to how and when the flexible adaption responses can be incorporated, WSCapp also identifies 'where' flexibility can be incorporated.

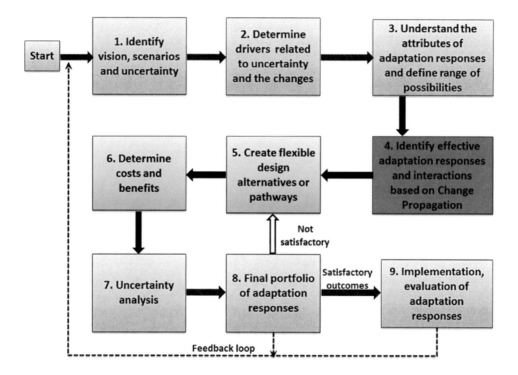

Figure 5-3 Flexible adaptation planning process for Water Sensitive city (WSCapp). The black arrows enclosed represent the sequence of steps in the WSCapp (clockwise). The dashed line from step 8 & 9 to step 1 represents the feedback to the vision cycle when a change in vision or objective happens over the time. The white arrow enclosed in black line from step 8 to step 5 represents the iteration with in the WSCapp, in case the final portfolio of measures obtained after the first run are not satisfactory.

The explanation of the individual steps in the flexible adaptation planning processes for a WSC, termed WSCapp, in an urban flooding context, developed from the previous discussions, is presented in Table 5-1.

Table 5-1 - Flexible adaptation planning process for water sensitive city (WSCapp) for incorporating and assessing flexibility in an urban flooding context

Required steps (Figure 3)	Description	Resources for additional Guidance on individual steps in WSCapp
1. Identify vision, scenarios and uncertainty	Ascertain what the visions for the city are, such as WSC, resilient and climate proof city based on ideas such as liveability, resilience and sustainability. Explore the possible scenarios in the future due to stressors such as climate change, socio-economics and the uncertainties associated with these.	Flood resilience in a water sensitive city context (Gersonius et al. 2016)

Transitions in a water management context (Ferguson et al. 2013a). |
| 2. Determine drivers related to uncertainty and the future changes anticipated | Methods such as adaptation pathways based on adaptation tipping points help in determining the impact of uncertainty on meeting the required objectives based on 'stress tests' in an urban area using physically based numerical models. Tipping points are the physical boundary conditions or the time at which the technical, economic, spatial or societal acceptable limits are exceeded (Haasnoot et al. 2011); i.e. when the system no longer provides the required service levels. | Stress testing and adaptation tipping points (Rodriguez et al. 2016) |
| 3. Understand attributes of WSC components and define range of possibilities | Flexibility attributes of the measure pertains to time, cost and effort required to change the scale, location and function of a measure. | Estimating annual flood damage Olesen et al. (Under Review).

Examples of attributes related to flexible design (Eckert et al. 2004; Spiller et al. 2015; Suh et al. 2007) |

Required steps (Figure 3)	Description	Resources for additional Guidance on individual steps in WSCapp
4. Identify critical WSC systems or components and interactions	Carry out relationship mapping between the adaptation measures. Determine the potential for change propagation in the urban system. Determine the potential for change propagation in the urban system: A change rarely occurs in isolation and impacts other systems. The adaptation responses that are capable of propagating more change (i.e. positive or negative impacts) when modified are potential adaptation responses for incorporating flexibility.	Mapping of relationships between adaptation measures - Infrastructure Victoria (Radhakrishnan et al. 2017a; Victoria 2016a; Victoria 2016b). Change propagation in complex engineering systems (Eckert et al. 2004) Flood resilience in water sensitive cities (Gersonius et al. 2016). Adaptation mainstreaming - an opportunity 'to adapt wherever we can, instead of wherever we have to' (Rijke et al. 2016).
5. Create flexible design alternatives or pathways	Adaptation pathways should be generated using the subset of adaptation measures based on relationship mapping and functionalities.	Adaptation pathways (Veerbeek et al. 2016) Pathway generator tool (Haasnoot and Van Deursen 2015)
6. Determine cost and benefit of design alternatives or pathways	The cost and benefits (also non-monetary) of implementation of adaption measures also includes the cost and benefits of switching from one measure to another in all possible scenarios.	Discounted cash flow analysis methods (e.g. Appendix B in De Neufville and Scholtes (2011)). Benefits of flood resilient adaptation measures - Gersonius et al. (2016) and BeST user manual (Horton et al. 2016).

Required steps (Figure 3)	Description	Resources for additional Guidance on individual steps in WSCapp
7. Uncertainty analysis	The present value or cost and benefit of pathways depends upon their performance in a scenario. As there are multiple future scenarios, the performance of pathways is likely to change leading to change in the tipping points.	Scenario planning for robust flood risk management (Brisley et al. 2016). Standard sensitivity analysis techniques - Monte Carlo Simulation (e.g. Appendix D in De Neufville and Scholtes (2011)) Model based performance analysis (e.g. Löwe et al. (2015))
8. Final portfolio of components defined and selected for WSC	Notwithstanding the benefits and costs outcomes, decision makers may decide to select a preferred strategy that is not cheaper in financial terms but may be their favoured strategy for other reasons.	Selection of preferred options - Gersonius et al (2016)

5.6 Practical considerations for applying WSCapp

WSCapp has been designed deliberately to resemble the steps in the contemporary adaptation planning process such as dynamic adaptive policy pathways and CRIDA steps in adaptive management cycle (CRIDA Under review; Haasnoot et al. 2013). This might lead to easier understanding and acceptance of WSCapp as an improvement of the existing adaptation planning processes, which is the overall aim of the research presented here. Hence, the emphasis in WSCapp is on the selection and assessment of flexible adaptation responses using the concept of change propagation (Step 4 in Figure 5-3). This is an important theoretical contribution to the adaptation planning process as WSCapp addresses the current gap on "where" flexibility can be incorporated. WCSapp will enable urban planners to analyse the impact of changes – such as change in architectural forms, open spaces and streetscapes – on liveability, sustainability and resilience. Also it is possible to visualise these changes due to urban design using outdoor augmented reality techniques (e.g. Calabrese and Baresi (2017).

Although it is not possible to accurately predict and model the changes in an urban environment, unlike in a factory setting, the results from the assessment can lead to informed decision making. There are examples and precedents in adapting concepts from other domains for planning and assessment of adaptation responses in urban adaptation. The concept of real options for valuing flexibility of adaptation responses (e.g. Zhang and Babovic (2012)) was adopted from valuing portfolios in stock markets. The agile urban adaptation technique that can be used for quickly responding to change and learning in an uncertain environment is based on software development techniques (Pathirana et al. 2017a). The assessment based on change propagation is both complex and computationally intensive as all the adaptation responses in various combinations in various scenarios have to be tested. However, once conceptualized, such computations are feasible with today's powerful computing techniques.

Some of the practical aspects that are be considered alongside the change propagation are the mainstreaming and the 5DA approaches that are used in adaptation planning (Gersonius et al. 2016; Rijke et al. 2016). Mainstreaming is suggested as an opportunity 'to adapt wherever we can, instead of wherever we have to' based on experience from cities such as Hamburg, Rotterdam, Malmo and New York (Rijke et al. 2016) (Table 5-1, Step 4 in Figure 5-3). Creatively embedding flexibility through mainstreaming by making modifications at the same time as other changes to the system may not increase upfront cost or the net cost, i.e., the cost of base elements together with the urban water element that is being mainstreamed with the base element. Also, mainstreaming is a collaborative effort where the requirements of multiple stakeholders are to be considered without compromising the interests of the asset owners. For example, a housing development project might give an opportunity to mainstream flood resilience in the form of blue-green infrastructure in courtyards. However, the risk is that the increase in natural ecological vitality provided by blue-green infrastructure may lead to a drainage facility becoming designated as protected against change to preserve the ecosystems, thus inhibiting the ability to adapt. Inspiration for resolving such impasses can be found from the manufacturing sector. For example, in the manufacturing of defence equipment such as helicopters there is a practise called 'offsetting' to avoid conflicts of interest of stakeholders (Eckert et al. 2004).

Offsetting is a practice where the components earmarked by clients are maintained (ring-fenced) even though they have the potential for flexibility. Instead, the manufacturer minimises the change propagated from earmarked (i.e., offset) systems by embedding flexibility or redundancy in the rest of the system that is not earmarked. For example, in a helicopter the radar and avionics systems are outdated quickly and have to be upgraded frequently to improve the performance of the helicopter. The manufacturer takes in to consideration the future upgrades – similar to car engine development – while finalising the overall design of helicopter. However certain buyers of helicopters such as military establishments many not prefer frequent changes because (i) it involves training their personnel to changing radar and avionics systems; (ii) there is a risk in not getting the upgrade done on time leading to reduction in serviceability of the helicopter; and, (iii) sanctions in future might hinder the procurement of the upgraded components in an ever changing political context. All these aforementioned factors affect the operational readiness of the aircraft, i.e., the capability to perform assigned flight missions (Verhoeff et al. 2015). Hence these buyers insist for a fixed radar-avionics configuration and prefer a design where flexibility can be incorporated in physical components which does not have the aforementioned drawbacks. A parallel can be drawn between offsetting in helicopters and urban flood risk management. For example an open area in a city can be converted into a wetland for detaining excess water during flooding and can eventually converted into a nature reserve. This conversion can lead to jurisdictional issues in the future between different city departments such as the estate department and parks and natural reserves department. Hence instead of converting the open area into a natural reserve the decision is offset, whereas an existing park is landscaped so that it can serve as a detention facility to hold excess water during heavy rainfall events. This arrangement prevents jurisdictional issues. This practice of offsetting is heavily context dependent and time dependent. Ascertaining the characteristics and the nature of flood resilience measures will help in identifying the measures that can be mainstreamed or need to be offset. For further details on offsetting in an urban context refer to Radhakrishnan et al. (2016)

Gersonius et al. (2016) recommend the 5 domains approach (5DA) to select flood resilience measures according to the nature of rainfall or stream discharge (see also Digman and Ashley (2014)). This includes the following related to environmental loadings: (i) day- to day events – potentially beneficial events which cause no damage; (ii) design events for which the system is designed according to set standards; (iii) exceedance events – which cause no or very little damage if managed effectively; (iv) extreme events – which cause substantial damages but within the recovery range; (v) unmanageable extreme events from which recovery is not possible. The 5DA classification is an improvement of the four domains approach – 4DA, Figure 5-4 (Digman and Ashley 2014). From Figure 5-4 it can it can be seen that the pressure and impact can be classified in to four domains based on the magnitude of impact and pressure the system is subjected to. The first three domains: (1) day to day domain; (2) design event domain; and (3) extreme domain with manageable damages are the domains where recovery is possible by means of resilience. Whereas, the unmanageable extreme events domain is beyond the recovery threshold and recovery is not possible. This classification can assist in ascertaining the effectiveness of the adaptation measure through determining the domain which the measure can be classified as the change propagated by the measure either within or across the domains. This can subsequently lead to the identification of where and how flexibility can be provided. For example, 5DA assessment in combination with change propagation assessment can be used to: (i) decide the locations where place making (e.g. water plaza) and landscaping roof terraces (e.g. green roofs) can be deployed, as they can efficiently cater to day to day and design events; and also to (ii) decide where the streetscapes (e.g. multi-functional streets) and open spaces (e.g. parks) can be modified and new architectural forms (e.g. amphibious dwellings) can be created as these responses can effectively cater to exceedance and extreme events. The aforementioned urban planning interventions / adaptation measures impact sustainability, resilience and liveability at varying levels and have a direct or "knock-on" effect depending on weather events such as heat waves, droughts or floods.

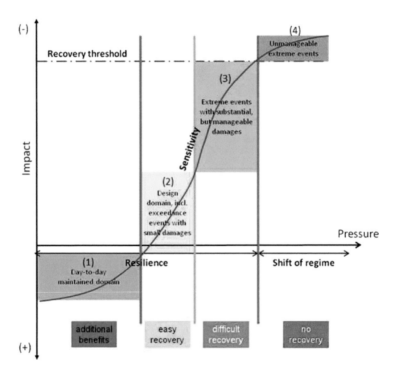

Figure 5-4 The 4 domains approach - 4DA recommends that classification of adaptation measures according to the type of loading, which the adaptation measure can address effectively (Digman and Ashley 2014)

Creativity in embedding flexibility is important as the best flexible designs when aligned with aspects such as mainstreaming or liveability, can increase the aesthetics and amenity value of the urban environment both now and in the future (e.g. Ashley et al. (In Press)).

5.7 Conclusions

This chapter sets out an approach (WSCapp) to identify where flexibility can be embedded in urban flood risk management systems as part of the essential components of a WSC. Knowledge and practices have been drawn from the domain of manufacturing industries, including automotive and aerospace, where tackling uncertainty using flexible designs is common. Comparisons have been made between the nature of issues and contexts of the automotive industry and the adaptation planning process in the context of a WSC. Flexibility can be incorporated into a WSC through identification of flexible adaptation responses based on change propagation, a regular practice in automobile assembly planning and design. Change propagation, which is the assessment of propagation of positive and negative impacts of adaptation responses such as SuDS and dewatering pumps, can be applied in an urban water management context. The Flexible adaptation planning process (WSCapp) based on change propagation can enhance adaptation in cities, where it is possible to identify and select adaptation responses such as rain water tanks and wet proofing of houses which lead to flexible as well as economic adaptation pathways adaptation responses (Radhakrishnan et al. Submitted c). Hence, WSCapp can be used by those who concerned with urban planning and urban adaptation to determine where the flexible adaptation responses can best implemented and to decide the nature of interventions such as enhancements to streetscape, place making or architectural forms that can contribute towards the objectives of a water sensitive city.

6 Flexible adaptation planning in a water sensitive Melbourne

"குணம்நாடிக் குற்றமும் நாடி அவற்றுள்
மிகைநாடி மிக்க கொளல்."

Verse 504 (Thiruvalluar 31 BC)

"Weigh well the good of each, his failings closely scan,
As these or those prevail, so estimate the man"
Translation based on (Pope et al. 1886)

Resilience towards climate and socio-economic change can be increased by means of flexible adaptation. In contemporary adaptation planning, building resilience is considered together with objectives such as sustainability, productivity and transformations. An adaptation planning process (termed WSCapp) may be used to incorporate flexibility or incremental flexible adaptation measures in a comprehensive adaptation strategy, such as when planning water sensitive cities. This paper has applied WSCapp in the context of adapting to urban flooding in Melbourne, which aspires to become a water sensitive city. Application of WSCapp has helped to identify appropriate adaptation measures; and economic adaptation pathways, which (in the case of Melbourne) included retrofitted drainage systems, rain water tanks at household level and the flood proofing of households. The application has shown that WSCapp can facilitate adaptation because it enables policy makers and planners to identify upfront the adaptation measures which are flexible and effective in a multiple objective adaptation context.

This chapter is based on the journal article *"Radhakrishnan M, Lowe R, Gersonious B, Ashley MR, Arnbjerg-Nielsen K, Pathirana A, Zevenbergen C (Submitted) Flexible adaptation planning in a water sensitive Melbourne Proceedings of the Institution of Civil Engineers - Engineering Sustainability"*

6.1 Introduction

The concept of a *"Water Sensitive City"* (WSC), i.e. a city being liveable, resilient, sustainable and productive whilst managing all aspects of the water cycle, is gaining popularity especially in developed economies (Howe and Mitchell 2011; Wong 2006). The WSC concept is becoming mainstream in Australia, whereas similar adaptation concepts such as transformative urban adaptation and resilient cities are gaining traction elsewhere in the world (EEA 2016; Revi et al. 2014; Spaans and Waterhout In Press). WSC and transformative urban adaptation concepts promote flexibility as an essential attribute to take advantage of opportunities from uncertainties. Flexibility in this context can be defined as there being opportunities arising from the number of alternative ways to provide services required when responding to changing circumstances (City of Melbourne 2016). Flexibility may also be seen as an essential characteristic of urban planning and infrastructure planning to deal with transformation in objectives such as becoming a water sensitive city (Ashley et al. 2013b).

Transitioning to a WSC needs a process, or processes that incorporate flexibility into planning, implementation and operation. An adaptation planning process for water sensitive cities is utilised here, known as WSCapp, to identify where flexibility can be incorporated into subsystems of a WSC, such as urban flood risk management systems (Radhakrishnan et al. Submitted-b). WSCapp has been developed drawing on knowledge and practices that are prevalent in the automobile and aerospace sectors, where adaptation – such as to changing customer requirements, technological advancements and market variations – is facilitated using flexible designs. This paper applies WSCapp to incorporate and evaluate flexibility into adaptation measures for managing flood risk in the Elster Creek catchment in Melbourne, Australia. The City of Melbourne was selected as a case study because, together with the State of Victoria it has begun to include flexibility explicitly in adaptation planning (City of Melbourne 2016; Victoria 2016a; Victoria 2016b).

6.2 Methodology

Adaptation and transformation in urban water management can be compared with the evolution of new car models in response to changing customer preferences and also technological advancements. For example, the consideration of aesthetics benefits of water in an urban landscape along with flood resilience in cities can be considered in a similar way to the preference for car engines with reduced emissions without compromising on the engine power and fuel efficiency. Cars not only provide transportation, but also offer the freedom to move and social status, which is similar to the concept of WSC, where water services, in addition to catering for basic needs, contributes towards enhancing the liveability and productivity in the city.

The automobile manufacturing sector uses product platform strategies, such as the flexible platform design process (Suh et al. 2007), to save costs by sharing core elements among different products in a product family (Figure 5-2). The uniqueness of this process is the identification of flexible components upfront to create an integrated platform, such as a car chassis, where the individual components can be changed easily in the future due to changing requirements (Suh et al. 2007). The incorporation of flexibility is based on the concept of change propagation, i.e., the components that are capable of propagating greatest change need to be assessed carefully before being selected as candidates for embedding flexibility (Eckert et al. 2004). According to the concept of change propagation, flexibility is incorporated in a location or in a component of the system that could minimise negative impacts and/ or maximise positive impacts when the system is subject to changing conditions (Eckert et al. 2004; Suh et al. 2007).

Similarly consideration of change propagation through systems such as rain gardens, rainwater harvesting tanks, mangroves – either due to the change in climate drivers such as rainfall, sea level rise and/ or change in vision or strategy such as from a water supply city to a water sensitive city – in an urban water context is essential to adapt in a flexible manner to changing circumstances. WSCapp (Figure 5-3) developed by Radhakrishnan et al. (Submitted-b) is based on the flexible platform design process developed by Suh et al. (2007).

The application of WSCapp follows steps 1 to 8 in Figure 5-3 indicated by the black arrow. The dashed line from step 8 to step 1 illustrates feedback, whereas the white arrow enclosed in black line represents the repeating nature of the analysis for each iteration where the decision makers do not favour the outcomes. Most of the steps in WSCapp are similar to the processes followed in the recent adaptation planning methods that are used in urban water management, such as real options, real-in-options, adaptation pathways and robust decision making (e.g. Gersonius et al. (2013), Haasnoot et al. (2012b), Zhang and Babovic (2012), Hall et al. (2012)). This similarity should help to facilitate the understanding and application of WSCapp. These methods make the overall planning and implementation process flexible. However, WSCapp has a unique aspect compared with these adaptation methods, as it focusses on the identification of effective adaptation measures where flexibility can be incorporated (Step 4 in Figure 5-3).

In a WSC, adaptation measures such as flood resilience measures are not only selected based on their ability to increase flood resilience, but also based on the ability to enhance the liveability and productivity. The identification of effective adaptation measures should be based on the following attributes: (i) Flexible or robust nature of the measure; (ii) secondary function of the measure; (iii) compatibility of the measures with other measures; (iv) change propagation in terms of resilience, liveability and productivity; and (v) inter-relationships between measures. For example, a conventional drainage system, although effective in reducing flooding, does not contribute to the liveability aspect of WSC, hence it is not an ideal candidate for incorporating flexibility. However, a rainwater harvesting system can also contribute to productivity in terms of reduced water consumption from the city network. Similarly, a rain garden can also enhance the aesthetics of the neighbourhood in addition to reducing the risk of pluvial flooding. Hence the change propagation aspect in a WSC context considers the change and the degree of change due to an adaptation measure in terms of flood resilience, liveability, and productivity.

6.3 Application of WSCapp in Elster creek, Melbourne

WSCapp has been applied in the context of adapting to flooding in the catchment of Elster Creek in Melbourne. Elster Creek`s coastal, low-lying area (i.e. Elwood in City of Port Phillip) is at the lowest point of a 40 km^2 urban river catchment and has been developed over drained marshland (Rogers et al. 2015). These characteristics mean that there is a significant flood risk which is predicted to increase with climate change due to frequent and intense rainfall events and rising sea levels. A combined 1D-2D hydraulic simulation in MIKE URBAN (Davidsen et al. 2017) for the different adaptation measures for various rain depths, sea levels and urban development states was undertaken to assess the flood risk. Simulated flood areas were intersected with land-use layers to compute flood damages using depth-damage functions (Olesen et al. 2017). The damages obtained from the different simulations were interpolated using a kriging approach to compute expected damages in different scenarios (Löwe et al. 2017). Long-term climate models have also projected hotter and drier conditions, presenting regional water scarcity challenges (CSIRO 2015). The steps in application of WSCapp and the relevance of its application in Elster creek are summarised in Table 6-1 with reference to Figure 5-3.

6.3.1 Identify vision, scenarios and uncertainty

Melbourne ranks highly among the most liveable cities in the world and aspires to become a resilient, water sensitive and business friendly city (Step 1, Figure 5-3). By considering the history of the changing visions and adaptation objectives in Melbourne it is apparent that these have changed from the protection of waterway health to that of a resilient water sensitive city, which is the present vision (Ferguson et al. 2013a). Urban resilience in the context of Melbourne is defined as the capacity of individuals, institutions and systems to survive and grow when exposed to chronic stresses and acute shocks. Effects of climate change and global trends such as urbanisation are evident in Melbourne and the City aims to increase its resilience (City of Melbourne 2016; Victoria 2014; Victoria 2016a).

Table 6-1 Application of WSCapp in Elster Creek, Melbourne

Flexible adaptation planning process in a Water Sensitive city context (WSCapp)	Relevance of WSCapp in Elster creek, Melbourne
1. Identify vision, scenarios and uncertainty	Melbourne now ranks highly among the most liveable cities worldwide and aspires to become resilient, water sensitive and business friendly. The City aims to increase its resilience through strategies such as survive, adapt, thrive and embed (City of Melbourne 2016; Victoria 2016a). Effects of climate change and global mega trends such as urbanisation are evident in Melbourne. The change in drivers of adaptation in Melbourne such as temperature, rainfall, sea level rise and urbanisation are uncertain (CSIRO 2015; Victoria 2014). Also there has been a continuous evolution in the vision and adaptation objectives within Melbourne (Ferguson et al. 2013a).
2. Determine drivers related to uncertainty and the changes	The main drivers of interest here are: Rainfall, sea level rise and urbanisation are drivers which are related to flood damages. However, the flood damages are especially sensitive and with a wide uncertainty range for rainfall and sea level rise in Elster creek (Olesen et al. 2017).

Flexible adaptation planning process in a Water Sensitive city context (WSCapp)	Relevance of WSCapp in Elster creek, Melbourne
3. Understand attributes of adaptation measures and define range of possibilities	In Elster creek, adaptation measures such as road elevation along the foreshore, lowering of street profiles, sea gates, foreshore mangroves, detention in parks and retrofitting existing drainage systems are being considered. These adaptation measures are effective in reducing the flooding; some of these measures contributing to liveability and have secondary benefits either as a detention facility or as a conduit in addition to their primary function. The maximum and minimum range of drivers such as rainfall and sea level changes within which each of these adaptation measures or their combinations that are effective are determined using physically based numerical models such as Mike Urban (Davidsen et al. 2017).
4. Identify effective adaptation measures and interactions	The range of adaptation measures whose attributes have been already understood by the planners or city managers and their performance ranges identified are the potential candidates for embedding flexibility. These measures are further subject to detailed analysis with respect to relationships with other measures change propagation, mainstreaming and offsetting. Offsetting is the practice of avoiding incorporation of flexibility in adaptation responses where there is a likelihood of operational constraints, ownership or jurisdiction issues and similar issues that involve multiple utilities (Eckert et al. 2004). In contrast, mainstreaming is actively looking for opportunities to implement adaptation measures together with other urban infrastructure components (Rijke et al. 2016). For example, the detention systems at household level such as rainwater harvesting tanks and at neighbourhood level such as parks, are measures that trigger major changes in the Elwood catchment. Increase or decrease in

Flexible adaptation planning process in a Water Sensitive city context (WSCapp)	Relevance of WSCapp in Elster creek, Melbourne
	household detention has a direct impact on detention volume to be provided in parks or the capacity of dewatering pumps. These are also the components where flexibility can be incorporated in case of scaling up, scaling down when changes are noticed in the trend of drivers of adaptation. It is relatively simpler to implement change in detention at household level or change the floor levels of properties undergoing renewal in response to the trends of sea level rise or rainfall instead of increasing dike height or making major changes to the pipe network.
5. Create flexible design alternatives or pathways	The adaptation pathways i.e. the combination of various adaptation measures for Elster creek were generated using an adaptation pathways approach and the tipping points were found to vary depending upon the pathways and individual climate scenarios. Tipping points are the points in time in future or predetermined values of variables such as sea level rise, rainfall, at which the objective of an adaptation strategy is no longer met or the required performance of an adaptation measure is compromised (Kwadijk et al. 2010).
6. Determine cost and benefit of design alternatives or pathways	The economic cost of individual adaptation measures was obtained from planning reports and from engineering firms. The present costs of adaptation pathways were obtained in relation to the tipping point of the measures in the pathway. The secondary benefits of water savings due to rainwater harvesting tanks was also considered.

Flexible adaptation planning process in a Water Sensitive city context (WSCapp)	Relevance of WSCapp in Elster creek, Melbourne
7. Uncertainty analysis	The uncertainty analysis of the present value of adaptation pathways for Elwood was undertaken for four climate change scenarios recommended by IPCC (IPCC 2013), CSIRO & BoM (CSIRO 2015). The present value of the adaptation pathways and results from the risk based approach were found to be sensitive to the changing climate and socio-economic scenarios.
8. Final portfolio of adaptation measures	IPCC scenarios are plausible scenarios and are assumed to have equal probabilities for calculating the expected present cost/benefits of pathways. Rainwater harvesting and flood proofing have the lowest expected present value of adaptation costs along pathways and in the risk based approach. Also these measures are recommended based on change propagation as rain water harvesting propagates positive change by reducing the peak flow of flood, whereas the household flood proofing reduces flood damages.

6.3.2 Determine drivers related to uncertainty and the changes anticipated

Elster creek an urban catchment in Melbourne, is subject to flooding and uncertainties related to likely increases in sea level, rainfall intensity and urbanisation (CSIRO 2015; Victoria 2014). The key drivers that affect the adaptation objectives were studied with the aid of numerical models and through stakeholder consultations (Step 2,Figure 5-3). The adaptation tipping point method helped in determining the impact of uncertainty in meeting the required objectives based on 'stress tests' by using numerical models (Rodriguez et al. 2016). Tipping points are the points in time in future or predetermined values of variables such as sea level rise, rainfall, at which the objective of an adaptation strategy is no longer met or the required performance of an adaptation measure is compromised (Kwadijk et al. 2010). Three drivers, being rainfall, sea level rise and urbanisation contribute to flood risk in Elster creek (Löwe et al., 2017).

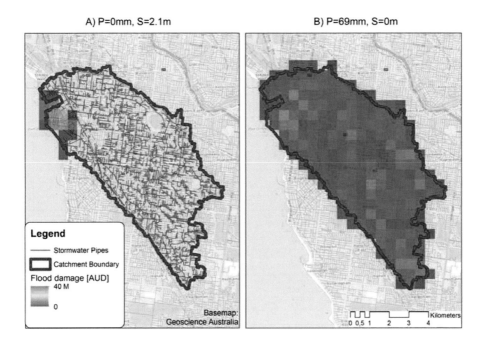

Figure 6-1 . Elster Creek catchment in Melbourne, Australia with flood damages simulated for 100 year return levels of sea level (A) and design rainfall depth (B) projected for 2090 in RCP4.5 (Löwe et al. 2017)

The range of uncertainty of climate drivers – rainfall and sea level rise for four representative concentration pathways (RCP) (i.e., as a result of possible mitigation actions taken by governments) – is as defined by IPCC (IPCC 2013). Under the RCP 2.6 the change in climate drivers such as temperature, rainfall and sea level rise will be the minimum, whereas under RCP 8.5, it will be the maximum. These two scenarios cover the entire range of variations. The estimated annual damages likely to occur due to coastal and pluvial flooding is depicted in Figure 6-1.

6.3.3 Understand the attributes of adaptation measures and define range of possibilities

Existing flood risk management plans for Elwood revealed a host of possible measures for addressing increasing flood risk due to sea level rise and higher rainfall intensities (Step 3, Figure 5-3). The adaptation measures have been compiled from various extant planning documents from the City of Port Phillip and Melbourne Water (e.g. Port Phillip adaptation pathways AECOM (2012), Flood management strategy Melbourne Water (2015), GHD (2014), Gunn and Rogers (2015)). The resilience strategy of Melbourne City (City of Melbourne 2016) emphasises the need for adaptation measures that can withstand chronic stress and also acute shock in all the scenarios. Modelling results show that measures such as road elevation, foreshore mangroves, lowering of street profiles, wet proofing of households, rain water harvesting and surface water detention at parks can withstand chronic stress and also acute shock for a wide range of rainfall intensities and sea level changes.

The analysis – based modelling results on uncertainties in sea level increase and rainfall, implementation costs and estimated annual flood damages in Elster creek– shows that the foreshore mangrove, rainwater harvesting, flood proofing of households, detention in parks and retrofitting can be the suitable candidates for incorporating flexibility. The net cost – across the scenarios – of elevating the road once using a robust design is less than (9.5 million Australian dollars) elevating the road twice using a flexible design (about 13 million Australian dollars), as the flexible design requires road formation twice that makes the design expensive. Hence, the road elevation measure can be a robust measure and is not suitable to be a flexible measure based on the cost and inconvenience incurred in embedding flexibility. Also with the exception of road elevation, all adaptation measures have a secondary function as an amenity in addition to their primary function. Also the adaptation measures in Elster creek are independent of each other, i.e., presence or absence of a measure does not hinder the performance of another measure. These measures are compatible with each other and compliment the functioning of each other. The attributes of adaptation measure are summarised in Table 6-2.

6.3.4 Identify effective adaptive measures and interactions

After determining the set of adaptation measures that are flexible, the change propagation aspects of those measures were determined (Step 4, Figure 5-3). The changes that propagate though the adaptation measures in the catchment pertains to flood resilience, liveability and productivity. For example, the extent of foreshore mangrove and its seasonality, i.e., presence, absence and the duration of the same is a subjective indicator of liveability (qualitative). The change in productivity was assessed based on the water saved due to the presence of rainwater tanks. Degree of change for individual measures and combinations of measures in varying proportions can also be determined for all scenarios.

6.3.4.1 Identification of effective adaptation measures based on change propagation

Detention at household level or at local level, using property level flood proofing measures is effective against flooding due to high rainfall events in Elster Creek. The detention of water or attenuation of peak discharges by household located measures propagates benefits throughout the system. Hence it was considered worthwhile to investigate in detail the changes propagated by these measures at household levels in the catchment, i.e. the reduction in flood damages. For example, rainwater harvesting can be mandated through local byelaws to detain a minimum amount of rainwater at households based on plot size.

Although most of the households in Elster creek have a standard $2m^3$ rainwater tank, a larger $5m^3$ rainwater tank was found to be effective in reducing downstream flood damages. The volume of detention can be revised in the future at stipulated intervals to reflect the changes in rainfall intensity over time.

Table 6-2 Attributes of adaptation measures in a Water sensitive Melbourne

Adaptation measure	Nature of the adaptation measure			Change propagation			Mainstreaming possibilities	Offsetting complications
	Robust / Flexible	Secondary function	Compatibility	Flood risk	Liveability	Productivity		
Road elevation	Robust	Yes	Yes	Yes	No	No	Yes	Yes
Foreshore mangrove	Flexible	Yes	Yes	Yes	Yes	No	Yes	Yes
Rainwater harvesting	Flexible	Yes	Yes	Yes	Yes	Yes	Yes	No
Wet proofing houses	Flexible	No	Yes	Yes	Yes	No	Yes	No
Drainage retrofitting	Flexible	No	Yes	Yes	No	No	Yes	Yes
Detention in parks	Flexible	Yes	Yes	Yes	Yes	No	Yes	No

Notes:

1. The nature of the adaptation measure, which is either robust or flexible is based on the cost incurred in making the design and implementation flexible

2. The primary function of all the measures is the reduction in flood risk, except road elevation where the primary function is connectivity and secondary function is reduction in flood risk. The secondary functions for other adaptation measure considered are ecological benefits, recreational benefits and economic benefits due to reduced water consumption.

3. The flood risk and change in flood risk, i.e., estimated annual damages is based on simulations, whereas the change in liveability and productivity are qualitative but can be computed.

4. The assessment of mainstreaming possibilities and offsetting complications are based on the present utility management practices prevalent in City of Port Phillip and Melbourne Water.

Similarly, there can be strict but revisable building regulations for minimum floor levels for houses that are at present under the flooding overlay levels; i.e., the special building overlay (SBO) of City of Port Phillip (CoPP 2016). There is a possibility of flood proofing when household assets are renewed (Nilubon et al. 2016). For example, if 4% of housing stocks come up for renewal every year, all the houses would have been renewed in 25 years. This is highly likely, as Melbourne is experiencing higher renewal rates due to rapid urbanisation (Victoria 2014). 25 years is also ample time to determine the increasing trend of rainfall intensities, based on which the regulation can be revised. Any change in the regulation regarding special building overlay or the preference for rainwater harvesting tanks is likely to have an effect on the liveability, productivity and flood resilient aspects of the catchment. This propagated change can be regulated using appropriate adaptation measures.

6.3.4.2 Identification of effective adaptation measures based on interrelationships

The other aspect that should be taken into consideration while selecting the component or subsystem for flexibility is the inter-relationships, i.e. the link between the adaptation measure with other measures or with stakeholders. These relationships may help in deciding where, how and when to implement the adaptation measure. The resilience plan for Melbourne proposes urban forestry as a flagship programme to promote resilience that also includes lowering flood risk and improving storm water quality (City of Melbourne 2016). The foreshore mangrove and upstream detention that has been identified as a measure for Elwood can be implemented under this urban forestry initiative. Improving the flood resilience of Elwood College can also be implemented through the neighbourhood plan that aims at training and building the community (City of Melbourne 2016). For example, the local council of Stawell in Victoria constructed a diversionary weir for flood mitigation, where the local population and businesses were involved in the construction through a community engagement programme (Seakins 2013). Similarly, the road elevation or street profile modifications are related to the drainage improvement as these measures are taken up in the same "right of way" of the streets and can benefit from each other.

Aligning adaptation measures together when there is an opportunity is also known as mainstreaming (Rijke et al. 2016). However, after identifying the effective adaptation measures, a thorough assessment of operational constraints, ownership or jurisdiction issues and similar issues that involve multiple utilities has to be undertaken before finalising the adaptation measures. This assessment process and finalisation of effective adaptation measures is known as offsetting, which is prevalent in the defence equipment manufacturing industry (Eckert et al. 2004). For example, the City of Port Phillip (CoPP) has the jurisdiction over the Moran reserve, a large area on the foreshore, whereas Melbourne Water is responsible for drainage of open spaces which have a surface area greater than 20 hectares. After converting the open space on the foreshore into a mangrove forest in collaboration with the Forestry department, it may subsequently be difficult for Melbourne Water and CoPP to intervene in future or to make other changes, as the legal status of the open green land will have become a "nature reserve" (Table 6-2). Similarly, the roads department may not easily agree to the change in road design that facilities the flow of water on surfaces or they might have a different renewal priority list of roads than the drainage authorities' list of flooded streets. In such instances coordinating adaptation actions will be complicated and the water authority could resort to offsetting. For example, the water authority could invest in high capacity dewatering pumps that could be moved anywhere in the catchment where flooding is anticipated. The strategy to invest in moveable dewatering pumps could also become a preferred option where the buy-in amongst the residents for a 'water sensitive city' way of living becomes less attractive and the City moves towards a utility based customer – service provider relationship between the residents and city council instead of the current position.

6.3.5 Creating flexible design alternatives or pathways

After identifying the individual flexible adaptation measures, the overall flexibility of the flood risk management system can also be increased by means of sequencing the Figure 5-3). An adaptation pathways approach can be used to generate the flexible adaptation pathways (Haasnoot et al. 2012b; Haasnoot and Van Deursen 2015). An adaptation pathway approach builds flexibility into decision making processes by sequencing a set of adaptation measures based on a 'tipping point' to changing circumstances in a range of plausible future conditions (Haasnoot et al. 2012b). The performance of the measures used or existing systems along the adaptation pathways and the tipping points – i.e., switching to another adaptation measure as there is a very high likelihood that adaptation objectives will be no longer met – were determined based on the expected annual damages. Out of the five flexible adaptation measures discussed in the previous section, three adaptation measures – (A) drainage improvements; (B) rainwater harvesting; and (C) flood proofing – have been considered to demonstrate the application of adaptation pathways.

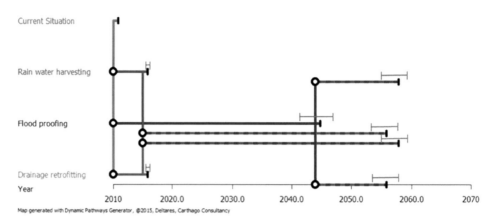

Map generated with Dynamic Pathways Generator, @2015, Deltares, Carthago Consultancy

Figure 6-2 Adaptation pathways, based on estimated annual damages (EAD), under Low climate scenario (RCP 2.6).Tipping point is reached when EAD increase to AUD 5.5million, i.e., 0.5 % of net annual revenue generated in Elster Creek. The black vertical line at the end of each pathway denotes the year (median value) at which the tipping point occurs. The range of tipping point is represented by the grey dimension lines shown above the pathways. Not all the pathways are shown here.

Rainwater harvesting and flood proofing at a household scale were selected as potential adaptation measures to support and manage change propagation (Table 6-2). Conventional drainage retrofitting measures were also considered as the base case; i.e. storage tanks and enlarged storm water drains. The estimated annual damages for each of these measures and for the combination of the measures has been calculated based on the year from which these measures were to be implemented. The expected annual damage cost (EAD) of AUD 5,500,000 was considered as the tipping point, which is equivalent to 0.5% of the net revenue generated in the Elster Creek catchment annually (Table 4, AECOM (2012)). A small EAD, i.e., tipping point, was selected with the intention to simulate the frequent tipping in order to demonstrate the performance of pathways with multiple adaptations measures within the planning horizon.

From Figure 6-2 it can be seen that the tipping point for the rainwater harvesting measure occurs at the year 2015 from a start date of 2010. However, when this measure is combined with flood proofing the tipping point is delayed and it does not occur until the year 2057. The tipping points of adaptation measures and pathways for four different climate scenarios based on CSIRO(2015) for Melbourne based on IPCC(2013) are presented in Table 6-3, whereas the pathways and tipping points are illustrated on Figure 6-2 (RCP 2.6) and Figure 6-3 (RCP 8.5).

From Table 6-3, Figure 6-2 and Figure 6-3 it can be seen that the tipping points vary depending upon the adaptation measures along the pathways and for the various climate scenarios. For example the tipping point of rainwater harvesting in households and drainage retrofitting occurs in the year 2015; whereas the tipping point of the flood proofing of households occurs between 2032 and 2044 dependent on the climate scenario. The rainwater harvesting measures and drainage retrofit are not as effective as the flood proofing measures in delaying the tipping point across the scenario. However, the combination of these measures postpones the tipping point. For example, when rain water harvesting and drainage retrofitting measures are combined along the pathway the tipping point of this pathway is likely to occur between the years 2038 and 2055 (Figure 6-2 and Figure 6-3). Hence, these combinations of measures are effective in delaying the tipping points at which service is no longer adequate.

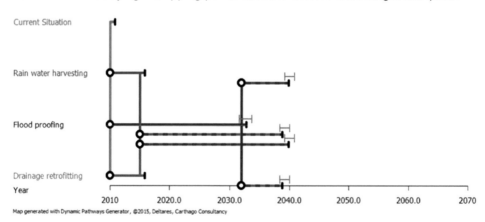

Figure 6-3 Adaptation pathways based on estimated annual damages (EAD), for extreme climate scenario (RCP 8.5). Tipping point is reached when EAD increases to AUD 5.5million, i.e., 0.5 % of net annual revenue generated in Elster Creek. The black vertical line at the end of each pathway denotes the year (median value) at which the tipping point occurs. The range of tipping point is represented by the grey dimension lines shown above the pathways. Not all the pathways are shown here.

6.3.6 Determine costs and benefits

The identification of preferred pathways can be based on an assessment of the cost and benefits that accrue along the pathways or the risks anticipated along the pathways (Step 6, Figure 5-3). The economic costs of individual adaptation measures were obtained from planning reports and from engineering consultancies (AECOM 2012; GHD 2014; Gunn and Rogers 2015; Melbourne Water 2015). The present costs of adaptation pathways discounted at 1.5% were obtained for the tipping point of the measures in the pathway (Table 6-3). The present value of the adaptation cost can also be calculated using a risk based approach (e.g. Kind (2014)), which can lead to a different set of adaptation measures, as the timing of implementation of adaptation responses are determined based on the reduction of overall risk during the entire planning horizon. Table 6-3 comprise the present value of accumulated total cost of flood damages and implementation cost of adaptation measures accumulated over 75 years. The uncertainty in the present value of the adaptation pathways for Elwood has been represented as a range, which comprises the present value of adaptation costs for the four climate change scenarios recommended by IPCC (IPCC 2013) and CSIRO & BoM (CSIRO 2015). The present value of the adaptation pathways and combinations was found to be sensitive to changing climate scenarios. Also from the benefit-cost ratio of adaptation measures (Table 6-3) it can be seen that a combination of measures, especially rainwater harvesting and flood proofing yields a better benefit-cost ratio across all the scenarios.

The rainwater harvesting measure has a bigger cost benefit ratio under RCP 2.6 as this measure is effective in reducing the flood damages only up to a certain increase in intensity of rainfall. When the increase in intensity of rainfall is higher, such us in scenarios RCP 4.5 to 8.5 there is no significant reduction in flood damages, consequently there is a low benefit cost ratio in these scenarios. The other reason for the higher benefit cost ratio, in general, for the rainwater harvesting measure is the low cost of rainwater harvesting compared with the flood proofing or drainage retrofitting measures. The rain water harvesting measure cost about 2300 AUD per household and the total cost of the measure is also offset by about 88 AUD due to the savings in water (Moglia et al. 2014). There are about 10,000 properties in Elster creek. The total cost of implementing rainwater harvesting is about 21 million AUD, whereas the cost of drainage retrofitting is about 996 million AUD.

Table 6-3 Tipping points and present value of adaptation costs for selective adaptation pathways in Elwood based on IPCC scenarios

Adaptation Measures	Tipping Point (median Year)				Present value of total cost of flood damages and implementation cost of adaptation measures accumulated over 75 years (in Million AUD)				Expected Value across scenarios (in Million AUD)	Investment Cost of adaptation measures (in Million AUD)	Benefit/ cost Ratio (Range , i.e., From RCP 2.6 to RCP 8.5)
	RCP 2.6	RCP 4.5	RCP 6.0	RCP 8.5	RCP 2.6	RCP 4.5	RCP 6.0	RCP 8.5			
No measure	2015	2015	2015	2015	5,444	7,984	8,119	15,872	9,355	-	-
A	2015	2015	2015	2015	2,042	2,506	2,538	4,103	2,797	996	3.4 to 11.8
B	2015	2015	2015	2015	1,413	1,944	1,980	3,742	2,270	196	20.6 to 61.9
C	2044	2038	2037	2032	655	986	992	2,242	1,219	180	26.6to 75.7
A & C	2055	2047	2047	2038	1,471	1,802	1,808	3,508	2,305	1176	3.4 to 10.5
C & A	2055	2047	2047	2038	1,456	1,686	1,694	2,586	1,856	1176	3.4 to 11.3
C & B	2057	2048	2048	2039	644	857	866	1707	1018	376	12.8 to 37.7
B & C	2057	2048	2048	2039	671	1,002	1,008	2,258	1,235	376	12.7 to 36.2

Adaptation Measures: A – Drainage retrofitting; B – Rainwater harvesting; C– Flood proofing households

6.3.7 Final portfolio of adaptation measures

The adaptation pathways can be assessed based on the present value cost across all plausible scenarios in order to select a preferred pathway (Step 8, Figure 5-3). As the IPCC scenarios are all equally plausible, they were assumed to have equal probabilities when calculating the expected present cost of the possible pathways or combinations thereof. Based on the lowest expected present value of adaptation costs and the highest benefit to cost ratio, the portfolio of flexible adaptation measures recommended for Elster Creek comprises rain water harvesting and flood proofing through the elevation of floor levels for households.

6.4 Discussion

In adaptation planning methods such as the adaptation pathways (Figure 6-2 and Figure 6-3) flexibility is realised by means of having a choice to implement or defer the adaptation measures. Here, the flexibility is a consequence of compatibility of the measure with the other measures in the adaptation portfolio. This facilitates the delaying or speeding up in terms of implementation based on the increasing rainfall intensities. However, the flexibility obtained through adaptation pathways is not due to the inherent inflexible nature of the measure i.e., structural or operational or functional flexibility(Radhakrishnan et al. 2016). WSCapp applied in the context of adapting to urban flooding in Elster Creek compliments the contemporary adaptation planning approaches. The significant outcome is that WSCapp has provided a structured approach to the identification of adaptation measures, such as rainwater harvesting and flood proofing in households based on the nature of the adaptation measures and the change propagated through these measures. The WSCapp also identifies the potential conflicts between the adaptation measures that might arise during the implementation and identification of measures that can be offset, such as drainage improvements and street profile changes.

Although the WSCapp, based on the flexible platform design process, helps to select the flexible adaptation measures, the role of the various stakeholders involved in the many process steps needs to be included. Also, the planning of adaptation actions must be considered in relation to the powerful and often ubiquitous political- economic interests in urban areas (Chu et al. 2017). The varying degrees of involvement and influence may be anticipated by the many and various players in the context of WSCapp applied in the Melbourne WSC context. For example in Elster Creek, the involvement of the National and Regional planning agencies would predominate in identifying visions and determining the drivers (step 1 and step 2 of Figure 5-3), whereas the role of the Melbourne regional planning authority would be predominant in setting the attributes of the WSC (step 3 in Figure 5-3). The role of the local council - City of Port Phillip - and waterway manager - Melbourne water - would be to lead in identifying the critical WSC components, creating flexible designs, calculating the additional benefits and undertaking the uncertainty analysis (steps 4, 5, 6 in Figure 5-3). Each of these various agencies can play an equal role in deciding the final portfolio of adaptation measures (step 8 in Figure 5-3). Hence in order to apply the WSCapp, effective stakeholder consultation, engagement and partnerships are necessary (e.g., agencies such as Municipal Association of Victoria can be engaged in the process of applying WSCapp in Elster Creek). Also, the application of WSCapp and identification of flexible adaptation measures early in the planning processes can improve the agility of the system; i.e., by establishing a system that can respond rapidly to uncertainty, threats and opportunities (Pathirana et al. 2017a). Although the change propagation can be quantified in terms of liveability, resilience and productivity, this paper has focused more on the aspect of flood resilience in order to demonstrate the application of WSCapp in Elster creek based on change propagation. This is a limitation of this paper and can be overcome with a comprehensive study which covers all the adaptation responses addressing the objectives of resilience, liveability and productivity.

6.5 Conclusions

In the context of enhancing urban resilience, flexibility is seen as being open to opportunities arising from the number of alternative ways to provide services required when responding to changing circumstances. Cities such as Melbourne are already incorporating flexibility to enhance the overall resilience of the city (City of Melbourne 2016). The flexible adaptation planning process formulated from synthesising the relevant literature and practice on flexibility incorporation and valuation has been demonstrated using the case example of Elster Creek catchment of Melbourne. From this case study it is evident that each of the steps of the WSCapp are relevant, needed and can be applied. It has been demonstrated that it is possible to identify the potential candidates for embedding flexibility in a WSC context. The next step would be to gather evidence during implementation– such as type of adaptation measures implemented, time of implementation of the measure, reasoning behind the selection – in order to strengthen the flexible adaptation planning process and increase the reliability of the approach further for the implementation of a resilience strategy. For example, evidence can be collected during the implementation of the City of Melbourne's overall resilience strategy or the State of Victoria`s infrastructure strategy to assess the limitations of the results presented here and WSCapp and address these before applying the approach more widely in Australia and elsewhere.

7 Operationalising flexibility: Agile urban planning process

காலாழ் களரில் நரியடும் கண்ணஞ்சா
வேலாள் முகத்த களிறு.

Verse 500 (Thiruvalluar 31 BC)

"The jackal slays, in miry paths of foot-betraying fen,

the elephant of fearless eye and tusks transfixing armed men"

Translation based on Pope et al. (1886)

Adaptation gaps are shortcomings of a system responding to climate change, whereas adaptation deficits are shortcomings in providing services. These two drivers for adaptation are often in conflict in many secondary cities in the global south (SCGS). It is possible to align these seemingly conflicting drivers into a productive unity, a conceptual alignment, which is the first step in achieving harmony while implementing adaptation actions. This paper focuses on the practical aspects of implementing aligned adaptation action that leads to improvements in liveability, sustainability and resilience of SCGS. At an abstract level, the nature of the adaptation problem is similar to the complex problems identified in various domains, such as software development, manufacturing and supply-chain management. The widely accepted 'agile principles' – used in the above domains – is the basis for developing a set of twelve principles for urban adaptation, which are synthesised from numerous recent studies that have implicitly proposed or applied most of these principles to climate change adaptation in urban settings. These principles lead into four essential objectives appertaining to the process of sustainable urban adaptation. The urban agile principles are used to analyse the current state of adaptation of Can Tho City in Vietnam and to ascertain the agile ways of addressing its adaptation challenges. Analysis of the outcomes show that harmonized approaches can simultaneously address both adaptation deficits and gaps.

This chapter is based on the journal article *"Pathirana A, Radhakrishnan M, Ashley R, Quan NH, Zevenbergen C (2017) Managing urban water systems with significant adaptation deficits – unified framework for secondary cities : Part II - conceptual framework Climatic Change"*

7.1 Urgency and uncertainty in urban adaptation

Cities have started addressing their needs to improve their liveability, sustainability and resilience. It is instructive to harmonize the often disconnected views on adaptation gap, adaptation deficit and the corresponding adaptation actions needed to close the gaps between these while addressing these needs of the cities. Adaptation deficit is the difference between the current state of a system and a state that minimizes adverse impacts from existing external (e.g. climate) and internal (e.g. urban development driven) forcings (Burton 2004; Pathirana et al. 2017b). The difference between any pre-agreed adaptation target (adaptation need) and the actual or anticipated state of adaptation is defined as the 'adaptation gap' (UNEP 2014). Predominantly, adaptation deficits look at the current state of affairs while the adaptation gap looks at the (medium and long-term) future (Burton 2004; Pathirana et al. 2017b).

Burton (2004) discusses two types of adaptation actions: Type I adaptation, the everyday adaptation to weather and climate that has always been a feature of human life; and Type II adaptation, the adaptation to (climate) change usually as mandated under the UN Framework Convention on Climate Change (UNFCCC). Type I adaptation is promoted as part of sustainable development, while Type II adaptation relates to anthropogenic climate change and is driven by rules and practices like those set out in the Convention. Hence it can be stated that Type I and Type II adaptation actions are those that are geared towards addressing adaptation deficits and adaptation gaps respectively.

Harmonizing Type I and Type II adaptation actions is important for all urban environments. However, it is imperative in the adaptation process of secondary cities, the rapidly developing cities and with a population less than five million, in the global south (SCGS), largely due to the existence of significant adaptation deficits. Pathirana et al. (2017)b have presented a conceptual framework – unified framework for adaptation – for harmonising Type I and Type II adaptation actions. The unified framework for adaptation brings together Type I and Type II adaptation actions in the context of (i) enhancing liveability in current situations; (ii) satisfying the demand due to short- term and medium term urban growth; (iii) ensuring sustainability of adaptation actions; and (iv) changing perceptions about adaptability among different stakeholders.

This chapter proposes a set of practical guidelines, recommend structures and processes to apply the framework into practice. The hypothesis presented in this chapter is that the 'agile principles' that are used in domains like software development to address the combined challenges of high uncertainty and high urgency can be utilised successfully to harmonize Type I and Type II adaptation of urban water systems. Agility in this context can be defined as the ability of the city to adapt quickly to the changes. The newly developed urban agile principles have been used to analyse the current state of adaptation of Can Tho City in Vietnam and to ascertain the agile ways of addressing its adaptation challenges. The relevance of the urban agile principles for SCGS in general is presented as synthesis and conclusion.

Adaptation in general, especially urban adaptation, has to address the three inter-related challenges, namely: a) uncertainty; b) path dependency3; and c) ensuring sustainability and liveability. These three challenges are closely related to addressing the adaptation gap (Pathirana et al. 2017b). However a fourth challenge arises while addressing the adaptation needs in SCGS, which is the need for strengthening basic infrastructure due to significant adaptation deficits. Significant deficits include: (i) lack of basic infrastructure (water supply, sanitation, roads, solid waste management etc.); and, (ii) lack of infrastructure to meet the increasing demand due the the growth of these cities, which increases the adaptation deficits over time.

The approach to adaptation implicitly embraced by decision makers and many stakeholders in SCGS is often sequential: (i) the adaptation deficit is addressed first by improving infrastructure; and (ii) once the deficit is reduced, adaptation gap is reduced. In other words, first address Type I adaptation and then move on to Type II. The sequential approach has two major weaknesses: a) Considering the significant

---

3 Path dependency is the causal dependence of a decision on past decisions. For example some types of Sustainable Drainage Systems (e.g. detention storage) will depend on the separation of storm water and sewage. Historical decision to have a combined sewer system has major implications on how these will be implemented in today's context.

adaptation deficits in SCGS (and the rapidly changing nature of stakeholder expectations), it is unlikely that they can be closed in a short time period; b) Important opportunities for addressing issues arising from uncertainty, path dependency and the need for sustainability are lost by following a sequential approach.

The worldview that is promoted by many scientists, even by international donors and central governments, is one that largely focuses on adaptation gaps related to climate change (e.g. IPCC (2014)). With justifiable scientific reasons, these key players attempt to promote the anticipation and addressing of medium and long-term consequences of global climate change. Understanding the need for climate adaptation research, many research donors rightfully make funds available for research in the domain. The sole promotion of the need to address adaptation-gaps often creates situations where the adaptation research ignores the adaptation-deficits (Type I) in contexts such as SCGS (e.g. UNEP (2014)). This can lead to the rejection or neglecting of Type II adaptation research outputs by authorities in-charge of urban development and many other locally relevant stakeholders.

In SCGS, a better approach is to consider Type I and Type II adaptation needs rather than to promote one or the other of these 'adaptation camps'. Integrative views are necessary to understand and appreciate any complex problem and to form realistic solutions and the management and adaptation of urban water systems is no exception (e.g. Allen et al. (2016)). Type I and Type II adaptation need to be integrated in projects in order to improve efficiency and sustainability. However, there are serious pitfalls to be avoided when following the path of integration in proposing and implementing adaptation responses. The pitfalls are (i) emergent complexity; and (ii) interconnectedness of the main and associated problems. City managers invariably favour 'transparent' information and struggle with complexity and the models that go with it (e.g. Hurley et al. (2008)). City managers, while appreciating the nuances of complex solutions, shy-away from them and move on with over-simplistic, traditional and sequential, approaches to address problems of what they see as 'real-world', i.e. they take a heuristic stance based on 'what-we-always-do'. In essence, the call for an integrative view often backfires, forcing city managers to largely treat the discourse as a theoretical one with little immediate practical value. Hence the challenge in applying the unified adaptation framework is ensuring that there is a practical plan of action that is (i) inclusive enough in-terms of representing reality; and,(ii) integrative. The plan of action should be straightforward and prescriptive enough to be of use to both the decision makers and practitioners– not a trivial feat.

We call here for the better alignment of Type I and Type II adaptation into a synergetic, cohesive and directed strategy, albeit with the caveat that implementation should not become 'lost' by being too well embedded; just as sustainability has virtually everywhere, and policy based climate changed adaptation in some cases (Chu et al. 2015). It is also recognised that the complete integration of climate adaptation issues with urban issues might lead to loss of momentum regarding tackling climate challenges and adequately dedicated attention (Chu et al. 2015). Hence the focus should be on harmonising the measures used to address Type I and Type II adaptation and not complete integration.

7.1.1 Addressing urgency and uncertainty in urban adaptation

Recent studies have emphasized the importance of inclusivity, integration and flexibility in the context of urban adaptation (Aerts et al. 2014; Allen et al. 2016; Birkmann et al. 2012; Chu et al. 2015; Garschagen 2014; Gersonius et al. 2012a; Klijn et al. 2015; Roberts 2014; Rogers et al. 2012; Spiller et al. 2015; UNEP 2014) For example, Gersonius et al. (2012c) emphasizes the importance of 'mainstreaming' adaptation action as an 'opportunistic' approach for implementation. Mainstreaming, properly applied, can aid in the harmonization of measures used for Type I and Type II adaptation. For example, suppose there is a major attempt to upgrade the sewer system of a city. This is a Type I adaptation action. Integrating flexible components that can be later upgraded as needed (e.g. Green infrastructure, such as swales) in response to future changes is a way to reduce path dependency in addressing future uncertainty (Ashley et al. 2016). Flexible components and portfolios of measures can provide a cost-effective way of sustaining the possibilities for addressing uncertain climate-induced pressures in the future. Hence this is usually a cost-effective way of integrating Type II adaptation with Type I actions.

The need to address adaptation deficits is usually urgent, but has only a moderate degree of associated uncertainty (Point B in Figure 7-1a) as current demands are relatively well known. Nonetheless projecting future from current demands always has a moderate degree of associated uncertainty. Whereas, adaptation gaps are characterized by a very high degree of uncertainty and a moderate degree of urgency (Point A in Figure 7-1a). Current demands to address these are small, but there are very high (*a priori* unknown) future demands with very high associated uncertainty. In SCGS, there is the need to successfully integrate or harmonise Type I and Type II adaptation measures in order to ensure the affordability, success and sustainability of the measures arising from adaptive actions. The combination is therefore a state that is both highly urgent and highly uncertain (Point C in Figure 7-1a). Integrated projects that address both Type I and Type II needs have a high urgency due to the large adaptation deficits to be addressed now and also high uncertainty due to the unknown future adaptation gaps.

7.2 Agile Principles

Addressing the combination of high uncertainty and high urgency is not a challenge that is novel or specific to the domain of urban water management. Similar challenges have been faced during the last several decades by, amongst others, manufacturing and Software development industries (Bernardes and Hanna 2009; Koste and Malhotra 1999; McGaughey 1999; Sánchez and Pérez 2005). Businesses and organizations such as these face a volatile environment that is highly uncertain with challenges such as increased competition, globalized markets, technology obsolescence and individual customer requirements. The uncertainties, the existential threat to survival and need for the maximisation of profits has necessitated an urgent shift of paradigm from traditional 'predictive' planning to a newer 'agile' or adaptive planning paradigm (Wendler 2013). This is illustrated in Fig 1b. Agility and agile manufacturing has been used for example by the US automotive industry to counter rigidity in manufacturing through the introduction of response buffers, postponing decisions in manufacturing and late configuring of products (Holweg 2005). Many new product industrial development software projects routinely apply agile methodologies that were originally formulated in the 1990s (Kettunen 2009).

The agility of an enterprise is defined as its ability to quickly respond to unexpected changes in order to survive unprecedented threats by proactively seizing opportunities. An agile approach (Figure 7-1c) may be illustrated using a number of principles, although the concept can be loosely defined as a process where requirements and solutions evolve through collaboration between self-organizing, cross-functional teams, promoted by adaptive planning, evolutionary development, early delivery, continuous improvement, and encouraging rapid and flexible responses to change (Fowler and Highsmith 2001; Leffingwell 2010). Therefore an agile process is one that has a high capacity to adapt to change (Pressman 2005).

The 'agile manifesto', that is prevalent in the software domain (Fowler and Highsmith 2001) and its twelve guiding principles such as continuous service delivery, welcoming change, maintaining simplicity, frequent updates, continuous engagement with stakeholders and sustainability presented in Table 7-1 have been derived from Beck et al. (2001)

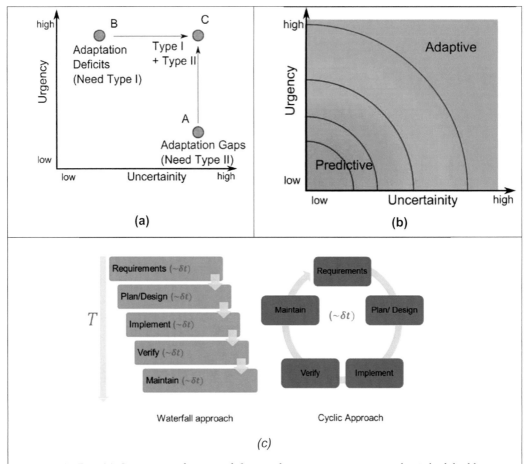

(a)

(b)

(c)

Figure 7-1 : Top: (a): Integrating adaptation deficits and gaps creates a situation that is both highly urgent and highly uncertain. (b) Moving from "predictive development" to "adaptive (agile) development" was the solution that worked in the domain of software development (after Chang (2010)). Bottom: Such shift of approach will also imply moving from a 'waterfall' approach to a cycle approach, where development is done in a number of, 'tight' cycles, each spanning a short amount of time/effort (in the order of $\delta t \ll T$ where T is the typical design horizon (say about 40-100 years for urban systems, and several years for software). δt can span between a few months to several years in case of infrastructure project (in software development it is typically a few days or weeks) Cyclic development allows for flexibility as well as learning by doing.

Table 7-1 Agile manifesto, agile principles for urban adaptation, essential objectives for agile urban adaptation and its relevance

Agile Manifesto for Software Development (Fowler and Highsmith (2001); Beck et.al (2001))	Relevant existing frameworks (Literature review)														Principles (Agile) for urban adaptation (Our interpretation)	Essential objectives for agile urban adaptation (What urban agile principles mean in practice)
	a	b	c	d	e	f	g	h	i	j	k	l	m	n		
1. Our highest priority is to satisfy the customer through early and continuous delivery of valuable software.	☑	☑	☑			☑	☑	☑	☑	☑	☑		☑	☑	1. Highest goal in urban development should be to enhance liveability, sustainability and resilience of urban communities, according to those communities' evolving aspirations.	A. Favour flexible, incremental, changeable, multi-value solutions a. While not always possible, make every attempt to propose adaptation measures that can be implemented incrementally – ideally at time-steps of several
2. Welcome changing requirements, even late in development. Agile processes harness change for the customer's competitive advantage.		☑				☑	☑				☑		☑		2. The aspirations of the stakeholders do (and should) evolve. Harness these changing aspirations and be ready to accommodate them, rather than attempting to resist them.	years. At every stage the solution should have utility. b. Favour solutions that result in as little path-dependency and lock-in as possible. c. Financial evaluation of measures should take into account the multiple-benefits of

Agile Manifesto for Software Development (Fowler and Highsmith (2001); Beck et.al (2001))	Relevant existing urban climate adaptation frameworks (Literature review)														Principles (Agile) for urban adaptation (Our interpretation)	Essential objectives for agile urban adaptation (What urban agile principles mean in practice)
	a	b	c	d	e	f	g	h	i	j	k	l	m	n		
																solutions, not only of the delivery of the main objective.
3. Continuous attention to technical excellence and good design enhances agility.			☑	☑	☑		☑	☑				☑			3. Consensus is not a license to do away with planning and engineering. Introduce planning and engineering to all the stakeholders. Focus on developing capacity of all stakeholders in the elements of planning and engineering necessary for meaningful contribution.	B. Virtual-worlds, understood and accepted by all stakeholders a. Comprehensive but easy to use modelling systems are essential. In the case of adaptation work, this is probably the only practical way to objectively test various proposed measures. The objective of the modelling systems is not to predict a single plausible future, but provide a platform to test variety of proposed measures
4. Simplicity--the art of maximizing the amount of work not done--is		☑						☑	☑	☑					4. Simple solutions are often the best. They are	

Agile Manifesto for Software Development (Fowler and Highsmith (2001); Beck et.al (2001))	Relevant existing urban climate adaptation frameworks (Literature review)														Principles (Agile) for urban adaptation (Our interpretation)	Essential objectives for agile urban adaptation (What urban agile principles mean in practice)
	a	b	c	d	e	f	g	h	i	j	k	l	m	n		
essential.															often easier to perceive, implement, manage and maintain.	against wide-range of scenarios. b. The modelling system should have trust, acceptance and agreement of all stakeholders. c. The modelling system should be routinely updated to incorporate new knowledge, and conditions, to address missing requirements. d. The modelling system should encourage experimentation with range of measures. It should never be a hindrance for testing measures. Sometimes it is needed to sacrifice complexity (resulting in slow response or user-unfriendliness) for utility.

Agile Manifesto for Software Development (Fowler and Highsmith (2001); Beck et.al (2001))	Relevant existing urban climate adaptation frameworks (Literature review)														Principles (Agile) for urban adaptation (Our interpretation)	Essential objectives for agile urban adaptation (What urban agile principles mean in practice)
	a	b	c	d	e	f	g	h	i	j	k	l	m	n		
5. Deliver working software frequently, from a couple of weeks to a couple of months, with a preference to the shorter timescale.		☑					☑						☑		☑	5. All adaptation solutions should have a value today, in order for them to be socially relevant.	C. **Type I and II adaptations are equally important.** a. Use today's requirements as starting points in evolving measures for addressing short, medium and long-term adaptation needs. b. Look for solutions that address both adaptation-deficits and adaptation-gaps. c. All solutions – even longer-term – should also have an immediate utility.
6. Working software is the primary measure of progress.			☑				☑	☑	☑		☑		☑	☑	☑	6. Never disengage the adaptation from regular urban development process. Look for opportunities for mainstreaming adaptation.	
7. Agile processes promote sustainable development. The sponsors,		☑	☑											☑		7. Done right, stakeholder groups and governance structures never become irrelevant, they	D. All stakeholders working and learning together: a. Promote leadership b. Organize city to city learning

Agile Manifesto for Software Development (Fowler and Highsmith (2001); Beck et.al (2001))	Relevant existing urban climate adaptation frameworks (Literature review)														Principles (Agile) for urban adaptation (Our interpretation)	Essential objectives for agile urban adaptation (What urban agile principles mean in practice)
	a	b	c	d	e	f	g	h	i	j	k	l	m	n		
developers, and users should be able to maintain a constant pace indefinitely.															should/can evolve and continue sustainably.	c. All stakeholders are important: Scientists are not first among equals . Stakeholders should also include users or clients of the system, e.g. citizen groups, business interests. Ideally these clients should be given the (informed – see f) leadership in the process of adaptation. d. Organize frequent stakeholder meetings. Physical meetings cannot always be replaced by teleconferences. In urban context this should be easy to facilitate. e. Use group-planning sessions
8. Business people and developers must work together daily throughout the project.		☑					☑			☑			☑		8. All stakeholder groups should participate in decision making from day one, all stakeholders should take shared ownership of decisions.	
9. Build projects around motivated individuals. Give them the environment and support they need, and trust them to get the job done.	☑	☑													9. Champions are indispensable: Identify them, empower them and help them flourish.	

Agile Manifesto for Software Development (Fowler and Highsmith (2001); Beck et.al (2001))	Relevant existing frameworks (Literature review)														Principles (Agile) for urban adaptation (Our interpretation)	Essential objectives for agile urban adaptation (What urban agile principles mean in practice)
	a	b	c	d	e	f	g	h	i	j	k	l	m	n		
10. The most efficient and effective method of conveying information to and within a development team is face-to-face conversation.	☑	☑		☑			☑								10. A community of practice of all stakeholders should meet physically, frequently. The collaboration needs a strong platform; it also needs objective means of testing and benchmarking diverse ideas against all conceivable scenarios.	effectively. Encourage all stakeholders (not only planners) to participate as equals. Use expertise of each constituent member only as a resource, not as a license for that member to monopolize the views related to the domain. f. Quickly fill the knowledge-gaps of all stakeholders by providing
11. The best architectures, requirements, and designs emerge from self-organizing teams.	☑		☑								☑				11. Properly guided, self-organizing stakeholder groups can often come with better solutions than a pre-prescribed plan. More importantly, they	appropriate learning opportunities (includes targeted trainings). g. Documentation of all important stakeholder activities is essential. Use easy-to-use modern

Agile Manifesto for Software Development (Fowler and Highsmith (2001); Beck et.al (2001))	Relevant existing urban climate adaptation frameworks (Literature review)														Principles (Agile) for urban adaptation (Our interpretation)	Essential objectives for agile urban adaptation (What urban agile principles mean in practice)
	a	b	c	d	e	f	g	h	i	j	k	l	m	n		
															encourage shared ownership of decisions.	collaboration platforms (e.g. wiki's) with version controls to achieve this. Try to make documentation an integral part of stakeholder activities
12. At regular intervals, the team reflects on how to become more effective, then tunes and adjusts its behaviour accordingly.	☑	☑									☑				12. Self-reflection of all stakeholders and the stakeholder group in-general helps in maintaining their relevance and effectiveness over time. City to city learning can be useful in this process.	

Relevant existing urban climate adaptation frameworks – review of existing literature on recent climate and urban adaptation

a Urban climate resilience framework (Tyler and Moench 2012)

b Inclusive approaches to urban climate adaptation planning and implementation in the global south (Chu et al. 2015)

c Urban futures methodology (Rogers et al. 2012)

d Methodology for local economic impact assessment of climate change (Hallegatte et al. 2011)

e Adaptive flood risk management planning (Klijn et al. 2015)

Agile Manifesto for Software Development (Fowler and Highsmith (2001); Beck et.al (2001))	Relevant existing urban climate adaptation frameworks (Literature review)														Principles (Agile) for urban adaptation (Our interpretation)	Essential objectives for agile urban adaptation (What urban agile principles mean in practice)
	a	b	c	d	e	f	g	h	i	j	k	l	m	n		

f Framing the adaptation Gap (UNEP 2014)

g Unified conceptual framework for Adaptation (Pathirana et al. 2017b)

h Design alternatives for flexibility (Spiller et al. 2015)

i Flood resilience strategies for coastal cities (Aerts et al. 2014)

j Framework for scope and scale of secondary cities (Roberts 2014)

k Integrated framework for vulnerability and adaptation analysis (Birkmann et al. 2012; Garschagen 2014)

l Flexible adaptive strategy (Gersonius et al. 2013; Gersonius et al. 2012c)

m Strategic flood risk management (Sayers et al. 2015)

n Urban infrastructure appraisal (Young and Hall 2015)

7.2.1 Agility principles in Urban Development

Recent literature setting out frameworks for adaption has (implicitly) included many of the proposed agile principles for urban adaptation. Table 7-1 also lists the components of agility found in the select urban adaptation frameworks reviewed. It can be seen that there is some congruity in each of the agility principles from the manufacturing sector when considered in the context of urban (climate) adaptation. As indicated in Table 7-1, many frameworks/methods have implicitly included a number of the principles. Chu et al. (2015) include most of these principles in the context of promoting public participation and facilitation of engagement of different civil society actors in adaptation planning and implementation. The inclusion of most components of the agility principles by Chu et al. (2015) is due to the inclusivity approach – as a means of stakeholder involvement and sustaining continuity in adaptation processes, based on a consideration of adaptation planning and implementation in two cities: Quito in Ecuador and Surat in India. These cities pursued two different inclusivity approaches such as broader inclusion and selective targeting of stakeholders respectively. Hence it has been concluded that the features of the agile approach advocated by the agility manifesto (Beck et al. 2001), are relevant to the domain of adaptation. The broad agile objectives of the manifesto that are applicable to agile urban adaption and considered in this paper are given below.

1. A strong emphasis on stakeholder's involvement: Processes and tools are secondary and are seen as a means of achieving proper stakeholder communication and representation of ideas.

2. Stakeholders do not have a 'blueprint' for the ideal product. Blue print evolves as the stakeholders collaborate through close, frequent communication and working in teams.

3. There is a 'working product' from the early stages of development (e.g. stakeholders do not have to wait for a distant future to witness the 'release' of a 'complete' product). The early prototype is, albeit a limited, but fully working system and is usable. This 'working state' is maintained throughout the subsequent development stages. Therefore there is no special emphasis on the 'final' product.

These features arise from embracing the complexity, uncertainty and urgency of the problem while at the same time using collaboration and respecting the needs and wishes of all stakeholders as a means of overcoming the challenges faced.

We propose here that a potential application of agile principles is in the context of urban adaptation: The conditions pertaining to the imperative uncertainty and urgency, Point C in Figure 7-1a, are similar between manufacturing needs and that of responding to climate and other urban system changes, hence the enabling mechanisms could be of a similar nature (Column 3 of Table 7-1). The potential for application of agile principles considered here is wider than the need to address urgency and uncertainty in urban adaptation for climate change alone.

On a cautionary note, the industrial systems such as automobile manufacturing, aerospace and nuclear energy systems which are "complicated systems", are different from urban water management systems, which are "complex" systems (Dunn et al. 2016). Application of complicated system principles such as agility principles to urban development needs therefore to align various components of the urban system and harmonise adaptation by understanding the known relationships of what is a complex urban system. The other area of distinction between the domains is in the scope of the outcomes between manufacturing and adapting urban areas. In the former, the aim is to produce goods and profit from sales, whereas, in a fair society, the aim of urban adaptation is to ensure the health, welfare and well-being of the urban population and environment. Municipalities typically oversee this task, although in recent times the private provision of public 'goods' has complicated this picture and profits may also be linked to the core services required of society, including urban water.

However, as application is in its infancy, these differences may not be a limitation to the use of manufacturing developed agile principles in the urban domain and will only be properly tested once there have been a number of applications. Also, it is useful to align or actively seek out the objectives of agile urban adaptation with the agile manufacturing frameworks in order to facilitate the application.

7.2.2 Essential objectives and practical principles for agile urban adaptation

Definitive quantification of various aspects of the medium and long term future of SCGS is not possible. The usual approach of aligning policy along a 'most-plausible' or 'most obvious' (i.e., business-as-usual e.g. Ashley et al. (2016)), single solution is not always useful. The adaptation challenge in SCGS is to reduce the lock-in or path dependency, which is due to the business-as-usual approach, as much as possible without excessive financial burden or loss of immediate utility. Further, the involvement of all the stakeholders, though complex, is likely to increase acceptability and sustainability of adaptation responses, contributing towards value creation.

Though it could be seen that the agility principles from manufacturing are valid with regard to urban adaptation and urban agile principles can be formulated (column 1-3, Table 7-1), it is essential to identify specific planning objectives in the urban adaptation context, that could be used in materialising / operationalising urban agility. The essential objectives of agile urban adaptation have been deduced from the existing urban adaptation frameworks (Refer column 2, Table 7-1) using agile manifesto and principles (Beck et al. 2001) as a guideline. The essential objectives of urban adaptation constitute: (i) flexibility – to tackle uncertainty and path dependency (Gersonius et al. 2013); (ii) plausible scenarios – to ensure liveability and sustainability under all potential futures (Brown et al. 2009); (iii) Type I and Type II adaptation - aligning and harmonising type I and type II adaptation (Burton 2004; Pathirana et al. 2017b); (iv) stakeholder consultations – views from all the stakeholders are paramount and continuous learning is achieved through engagement at every state of adaptation, such as in learning and action alliances (Ashley et al. 2012; Cettner et al. 2014a) . The collection of practical objectives should encompass all of the requirements expressed by the twelve agility principles (Refer to the grouping of agile urban principles based on the essential urban objectives in column 4, Table 7-1). These outline what should be the essential objectives of a viable urban adaptation system.

The prevalence of agility principles or the four agile objectives (A-D in column 4 of Table 7-1) in the current urban adaptation context is most apparent when considering SCGS. As an example, Can Tho city in Vietnam is undergoing climate and socio-economic adaptation and has been selected as a case study city to review the application of agility principles and identification of agile urban adaptation objectives.

The various adaptation aspects in Can Tho have been assessed using the principles and the four objectives of a proposed agile adaptation framework. Can Tho case study should not be seen in the view of operationalising or applying the agility principles in adaptation planning. The case study illustrates where the agile objectives are identified in an ongoing adaptation planning process. The implementation issues and other practical difficulties in operationalising agility principles in urban adaptation is work under progress and will be reported in a follow up article.

7.3 Possibilities for applying agile principles in Can Tho

Can Tho is an SCGS and the largest and a fast growing secondary city in the Mekong Delta, Vietnam. The city is affected by drivers such as (i) sea level rise, driven by climate change; (ii) change in river runoff, due to climate and hydrologic change; (iii) change in urban runoff, due to imperviousness growth and enhancement of extreme rainfall due to urban growth driven micro climate; (iv) socio-economic drivers such as population growth, spatial planning and economic growth rates (Huong and Pathirana 2013; MDP 2013; SCE 2013). However, there is uncertainty with regard to climate and socio economic drivers. Assessment of proposed urban development and adaptation responses –such as new ring dikes, improvements to roads – show that the adaptation responses, which are path dependent and address only the current adaptation deficit, are not sustainable in the face of swift and uncertain future change (SCE 2013). Can Tho was selected as the case study city as it is a secondary city with Type I and Type II adaptation needs and the authors have a good understanding of adaptation needs as they have been working with this city on various adaptation projects.

Increased in-migration and unaffordable real estate prices have meant that a section of the population has settled alongside rivers and canals thereby making themselves vulnerable to pluvial and fluvial flooding due to both an infrastructure deficit and change in climate (Garschagen 2014; Quan et al. 2014). Can Tho is adapting to climate change as well as the socio-economic changes and to a certain extent is still adapting to the political changes after the Indochina wars, a series of wars fought in the region between the years 1946 to 1989 (Garschagen 2014; World Bank 2014). Hence Can Tho provides an opportunity to review the application of the agility principles for urban adaptation. We have evaluated the current status of adaptation planning and processes in Can Tho against the four main objectives of the proposed agility framework (Table 7-1). In the following discussion each of the twelve agile principles listed in Table 7-1 are referred to as '[Px]' where x is the number given to the principle in Table 7-1. The relevance of agile urban principles and agile urban objectives in Can Tho are summarised in Table 7-2.

7.3.1 Are there generally accepted 'virtual-worlds'?

Studies on Can Tho residents' vulnerabilities, adaptation to climate change (Garschagen 2014) and proposed dike system (Pham et al. 2009) are available. Radhakrishnan et al. (2017b) for example, combine global and local understandings of climate change and socio economic change to estimate the enhancements to adaptation pathways for the planned flood protection systems in Can Tho due to enhancements in the coping capacity of the citizens. The recent Mekong Delta Plan presents four different regional development paths based on climate change, urban development and industrialization (MDP 2013). There are a number of modelling studies of varying degrees of complexity and integration (Huong and Pathirana 2013; Quan et al. 2014; Radhakrishnan et al. 2017b), although there is no general acceptance among stakeholders which models ('virtual worlds') to use to explore the various scenarios (Table 7-2). The lack of an accepted set of 'virtual worlds' has resulted in a lack of common understanding of the various adaption issues and outcomes among different stakeholders [P3]. While attempts are being made in the context of a number of training and information sharing initiatives by different projects, considerable work has yet to be done to clarify both the issues and the potential responses available in Can Tho.

7.3.2 Equal importance for Type I and Type II adaptation

Some of the current adaptation measures such as dike heightening to overcome the adaptation gap in Can Tho have been found to be effective only for a short time into the future (Radhakrishnan et al. 2017b). Many of the current and planned Type I and Type II adaptation measures for climate change and socio economic change are also not compatible (Pathirana et al. 2017b). There are proposals (PM 2013) for the socio-economic development and poverty reduction of Can Tho which project the plans as inclusive, pro-poor and delivering social justice. Measures such as: (i) The relocation of vulnerable households to reduce exposure to floods – which are not effective in Can Tho as residents will not move as there is a fear of loss of income due to the relocation; (ii) The possibility of elevating the houses in-situ – which depends not only on needs due to past occurrences of flooding but also on the economics of individual households and any savings available for self-financing (Garschagen 2014). The type II adaptation measures such as dike heightening that are being planned are based on the historical river water levels along with allowances for climate forcings such as sea level rise (eg. SCE (2013)). These do not account adequately for uncertainties and are not flexible in nature and therefore are of high risk. Type I and II adaptation measures are also being promoted separately by different stakeholders and interest groups and there are major differences of opinion amongst these on the relative importance of the various measures proposed (Table 2). Hence the present value of adaption action is not well reflected in the adaptation plans [P5] and equal importance is not given to Type I and Type II adaptation [P6].

7.3.3 Lack of flexibility and value from the proposed adaptation measures

The current measures addressing Type I and II adaptation in Can Tho are not flexible (Pathirana et al. 2017b; Radhakrishnan et al. 2017b). However, there is scope for incorporating flexibility measures in dikes to tackle uncertainty against sea level rise and how this could be achieved has been demonstrated for Can Tho (Radhakrishnan et al. 2017b). Looking at these measures from a flexibility point of view can open opportunities for simultaneously achieving type I and type II adaptation goals as well as securing other multiple benefits [P1]. There is a considerable scope for improving the potential (macro-) economic benefits of adaptation projects by looking at the opportunities for multiple benefits [P2]. Currently this is an area which is largely neglected in Can Tho (as in many other SCGS). For example, Quan et al. (2014) demonstrated that there is scope for addressing adaptation gaps in an urban development project with the dual objectives of poverty alleviation and beautification around the Xan Thoi lake (Table 7-2). However, the necessary alignment of adaptation actions for this requires sustained involvement of stakeholders and a general methodology that accepts and includes engagement from all the stakeholders at all stages of adaptation, e.g. using learning and action alliances (cf. Ashley et al. (2012)).

Table 7-2 : Scoping for urban agile principles and identification of agile adaption objectives in Can Tho

Essential agile objectives in Can Tho (Refer to Table 7-1 for description on objectives and principles)	Urban agility principles needed in Can Tho
The current adaptation measures that are being planned and implemented in Can Tho are not flexible and do not have multiple value, but there is scope for flexibility and multiple benefits (Urban agile objective A - Favour flexible, incremental, changeable, multi-value solutions- comprising principles P1, P2)	The current proposals address the resilience component and liveability component separately and not together. The sustainability components are not clearly reflected in the adaptation and urban development plans.
	The change in aspirations of people are not reflected in the plan but there is scope for inclusion.
In Can Tho, it can be seen that there are multiple virtual worlds - various engineering, planning and socio economic models - but there is a lack of common understanding. (Urban agile objective B - Virtual-worlds, understood and accepted by all	Periodic attention is paid towards technical excellence of design standard such as flood protection (e.g. SCE 2013) However, the efforts do not encompass all key stakeholders
	Many solutions are sector-based, making them not-simple, but complex to maintain and integrate with overall development of the city

Essential agile objectives in Can Tho (Refer to Table 7-1 for description on objectives and principles)	Urban agility principles needed in Can Tho
stakeholders - comprising principles P3, P4)	
Type I and Type II adaptation needs are clearly seen in Can Tho, although the focus there seems to be more on Type II adaptation. This could be due to the interest or agenda of international donors in Can Tho. (Urban agile objective C - Type I and II adaptations are equally important - comprising principles P5, P6)	Present benefits of future adaptation requirements is not clear in the adaptation plans proposed. Most of these plans talk about present costs and future benefits.
	Satisfying the current urban development needs through adaptation plans is not being done, although there is scope for this (e.g. Quan et al. (2014))
Although a number of stakeholders are involved in the adaption planning of Can Tho, there has not been any significant effort put towards learning and working together. There is plenty of scope for it and it is being explored	Can Tho is still undergoing evolution in governance due to the change in status from a city to a city state. The aspirations of the people are also evolving as seen in the rise of middle class and in the perceptions' about adapting to floods. Refer to Garschagen (2014)
	Although there is interest among the diverse group of participants towards participation and to take ownership of decisions, there are no platforms to do this and there are social barriers hindering equal participation.

Essential agile objectives in Can Tho (Refer to Table 7-1 for description on objectives and principles)	Urban agility principles needed in Can Tho
through capacity building workshops and city to city learning networks. (Urban agile objective D - All stakeholders working and learning together - comprising principles P7, P8,P9,P10,P11,P12)	Involvement of champions in adaptation issues are not clearly evident in Can Tho. However, the need for champions and their involvement is being encouraged through multi-stakeholder workshops (e.g. Radhakrishnan (2015)).
	Meeting of all stakeholders is not yet common. When this happens, the interactions can be too rigidly controlled (Clemens et al. 2014)
	Although there is scope for emergence of self-organising stakeholder groups and consultations, it has to be nurtured and guide in the initial states
	The relevance of present and future adaptation plan in Can Tho is yet to emerge through stakeholder consultants although there is scope for this.

7.3.4 Stakeholders working together

A number of stakeholders such as the Department of Planning and Infrastructure (DPI), Department of Labour, Invalids and Social Affairs (DOLISA), Ministry of Agriculture and Rural Development (MARD), Ministry of Natural Resources and Environment (MONRE), Can Tho Climate Change and Coordination Office (CCCO) are working on aspects that are addressing the adaptation deficits and gaps in Can Tho i.e., [P7] and [P8] (Clemens et al. 2014; Garschagen 2014). The organizational setup at the higher levels of government and at grassroots level seems to provide a means for involvement via stakeholder meetings, the promotion of leadership and a flow of information at some levels (Clemens et al. 2014). However, there is rarely any common understanding or agreement between these stakeholder communities about the adaptation issues such as: i) the acknowledgement that climate adaptation and socio economic adaptation are occurring in parallel, resulting in uncoordinated planning and implementation by DOLISA, MONRE and MARD; ii) adaptation is an environmental issue that may conflict with economic development, which subsequently results in protectionism separating the perspectives of MONRE and MARD (Garschagen 2014); iii) local communities feeling that lower-tier stakeholder aspirations and options do not reach the higher levels of government (Garschagen 2014; Moench et al. 2011). Numerous research projects have organized seminars and workshops with representation from other cities within Vietnam and overseas, so that the city managers could learn from the global experience. Encouragingly, the ideas expressed in 'City to city learning' (Zevenbergen et al. 2015b) are gaining acceptance in Can Tho. From the above narrative it is clearly evident that [P9], [P10] and [P11] are not being included effectively in Can Tho (Table 7-2).

7.4 Discussion

It is clear that many of the objectives necessary for effective agile adaptation, such as a common understanding of adaptation, mechanisms for stakeholder involvement and champions are not yet available (Table 7-2) in Can Tho. Many SCGS – such as Byblos in Lebanon, Porto Alegre in Brazil, Semarang in Indonesia – face the same situation (100 Resilient Cities 2017). This study has identified that Can Tho has the potential for developing an agile, learning and dynamic stakeholder community and many of the key components for this such as Type I and II adaptation needs, adaptation plans with potential for flexibility, aspirations among stakeholders to cooperate are in place (Table 7-2). The current challenges

are to: (a) create an enabling institutional and governance environment that promotes these potentials and: (b) integrate the various elements together in a viable fashion.

Can tho city has significant capacity gaps in terms of adaptation. Although there is an understanding of climate issues, there is a lack of shared knowledge about adaptation measures, interest and capacity for learning. There are few if any, practical avenues for knowledge pooling and discussion although there is a central coordinating office for climate change CCCO (Garschagen 2014; Moench et al. 2011). There are numerous initiatives often associated with research projects, with the objective of enabling information sharing, collaboration and promoting inter-sectoral coordination. However, these attempts are diffuse and are yet to reach the critical mass and inertia needed to be effective.

The analysis presented here has revealed the benefits of applying the agility principles to understand the limitations of the adaptation landscape in Can Tho as a typical SCGS. The agile adaptation principles can be applied to align adaptation measures, harmonize multiple needs, and bring about organisational integration and synchronization at multiple levels. This includes households, district, city, provincial and national levels and coordination across agencies such as DOLISA, MARD and MONRE to better understand, plan and implement adaptation measures.

Although agility principles are demonstrated here as being applicable in the context of SCGS in Can Tho, these are likely to be applicable in a city in a developed or industrialised country as such cities are also undergoing significant change in the context of increasing resilience, sustainability, liveability, transformation and productivity. Rapid change can create situations with large adaptation deficits even in these cities.

An argument against the application of agility principles is the observation that urban adaptation context is far more complicated compared with the original domains the agility principles were borrowed from. Urban adaptation plans are: (i) often outdated by the time they are ready to be implemented; either due to adaptation deficits or changes in external stressors (Pathirana et al. 2017b; Radhakrishnan et al. 2017b); (ii) susceptible to manipulation by political decisions and the purchaser – provider relationship between e.g. international development banks and national governments (Cettner et al. 2014a; Poustie et al. 2016). However, in spite of these added challenges, adaptation in urban areas needs to: address path dependencies, adhere to principles of sustainability and importantly, pursue the improvement of liveability by addressing adaptation deficits. Urban adaptation needs include

the essential traits of agile systems such as the changing dynamics between the system components and the continuously evolving ways in which the objectives such as liveability are addressed. Hence application of agile principles is recommended as a useful framework for formulating urban adaptation responses.

Transformational adaptation is the way of using behaviour and technology to change biophysical, social or economic components of a system fundamentally (EEA 2016). Urban adaptation provides opportunities for incremental and transformative development trajectories, where there is an integration of knowledge into decision making, building on exchange among policy makers, scientists and those at risk; designed and modified to the local needs and capacities; ensuring plan and policy continuity; promoting ownership and equity (Chu et al. 2015; Revi et al. 2014). The application of agile principles to urban adaptation is proposed here as a practical way to operationalize transformational urban adaptation.

7.5 Conclusion

Approaches to adaptation to climate and societal changes are typically fragmented into two often not connected types: Type I - adaptation deficits; Type II - adaptation gaps. This fragmentation is inefficient, likely to lead to extra costs and potentially ineffective responses. Urban adaptation is necessarily a complex problem exemplified by the combined realities of high uncertainty and extreme urgency to act. At the same time the pitfall of inaction due to unmanageable complexity needs to be avoided. The traditional predictive approach of understanding the problem and then designing and planning solutions for the long-term often fail in SCGS. We believe that a set of strategies similar to the proposed framework of using agility needs to be utilized to harmonise urban adaptation. Many fields that are far removed from urban adaptation have faced similar challenges of having to address urgent, but poorly-understood problems. Here we have utilised the experiences from information technology and automotive sector in resolving uncertain and urgent urban adaptation challenges.

Addressing the various adaptation needs by harmonizing processes for synchronisation of the ways of dealing with adaptation deficits and gaps is essential, especially in the Secondary Cities in the Global South (SCGS). Pathirana et al. (2017b) set out a proposed conceptual framework to address the challenge of harmonizing these adaptation deficits and gaps whereas in the current paper a set of agile principles and objectives are set out that can help identify and address shortcomings in the adaptation responses of urban areas. This can direct efforts to harmonize approaches to simultaneously address both adaptation deficits and gaps. By adapting 'agile' approaches the challenges exemplified by the combination of high-uncertainty/high-urgency has been addressed here by encompassing the complexity arising from inclusion of multiple players, drivers and scenarios. Based on the well-established 'twelve principles of agile software development', we have proposed an equivalent set of principles for urban adaptation (Table 7-1). Review of selected recent literature related to urban adaptation shows that similar ideas are being promoted in many discussions on urban adaptation (Table 7-1).

The established principles of agility have been interpreted into twelve guiding principles for application in adaptation planning. Further, the four important objectives of urban adaptation that are proposed – flexible incremental solutions; common understanding (i.e. through virtual worlds); equal importance to adaptation gaps and adaptation deficits; and stakeholders working together - that will help bring these principles into action. It has been demonstrated in a case study of adaptation in Can Tho, Vietnam that the agility approach, derived from manufacturing and software industries, is applicable to urban area adaptation. The principles and objectives set out here should only be seen as the first steps towards a verifiable approach that can only be achieved from further application and experience.

8 Conclusions

"வெள்ளத் தனைய மலர்நீட்டம் மாந்தர்தம்
உள்ளத் தனையது உயர்வு"

Verse 595 (Thiruvalluar 31 BC)

"With rising flood the rising lotus flower its stem unwinds;
The dignity of men is measured by their minds."

Translation based on Pope et al. (1886)

Flexible adaptation measures enhances flood resilience. The aim of the research was to explore approaches to maximise the effect of flexible adaptation practices towards achieving the desired adaptation outcomes in an urban environment by means of structuring the various adaptation outcomes and through the identification of flexible adaptation components. Hence this research has focused on increasing the knowledge and support the practice of incorporating flexibility into urban flood risk management systems, in order to facilitate adaptation in the context of climate change. The application of adaptation planning processes in flood risk management such as adaptation pathways, real options and real-in-options were reviewed (Buurman and Babovic 2016; Gersonius et al. 2013; Haasnoot et al. 2013; Haasnoot et al. 2012b; Woodward et al. 2014; Zhang and Babovic 2012). Also, the evolution in adaptation planning and implementation, for example, to transformational adaptation was also reviewed (e.g.EEA (2016), Wong and Brown (2009)).

From these reviews, it is concluded that there is (i) lack of integration of objectives (across disciplines), which calls for a multiple perspective integration; (ii) lack of spatial integration (across spatial scales), which calls for a system approach (a system is composed of sub systems and they are all interacting), recognition of emergent behaviour (emergence) and synergistic effects; and (iii) lack of temporal integration (across time scales) dealing with uncertainty, short term connecting to long term etc., which calls for a flexible approach and learning Figure 8-1. This lack of understanding and ignorance is prevalent at the higher planning level, such as regional and country level, and also at the local level of planning and implementation, such as city level.

At a higher planning level there is lack of understanding of the relationship between the adaptation measures and adaptation drivers, and also the specific adaptation needs at a local level. Whereas, at the local level there is a lack of understanding about the future trend of global adaptation drivers such as climate change. Irrespective of the level at which adaptation planning is made adaptation responses are considered as standalone interventions and not as an integral part of the project. Further, flexibility is used only as a means to postpone these capital intensive standalone interventions to enhance the implementation of no regrets or small low cost investments. Also majority of these decisions are based on economic criteria.

The current adaptation planning processes that are often centred around a single criteria has been guiding the adaptation research towards maximising efficiency in achieving the desired adaptation objectives, which are mostly economic.   Hence there is a dearth for adaptation research on implementation that can explore the processes and enabling mechanisms in order to make the implementation of adaptation responses agile and inclusive. Such implementation processes should also induce learning – such as what went right and what went wrong during adaptation – and allow for experimentation during the implantation of adaptation responses. The pursuit for developing effective adaptation implementation processes will also create opportunities to learn from sectors, for example aerospace, automobile industry and health care, which are not related to climate or urban adaptation. The observations during the research, findings, actions, outputs and anticipated outcomes are summarised in Figure 8-1.

The observations of this research and the course of this research are similar to the fourth generation adaptation research that focus on adaptation implementation (Klein et al. 2017). A comprehensive understanding of adaption measures and relationships between them are prerequisites to deciding on measures, approaches, pathways to maximise the value of flexibility in adaptation planning and implementation. This led to the identification of a gap in knowledge about the process, i.e., lack of comprehensive understanding by the researchers and the stakeholders such as planners, policymakers and implementation agencies of local adaptation responses and an absence of frameworks to structure the adaptation responses.

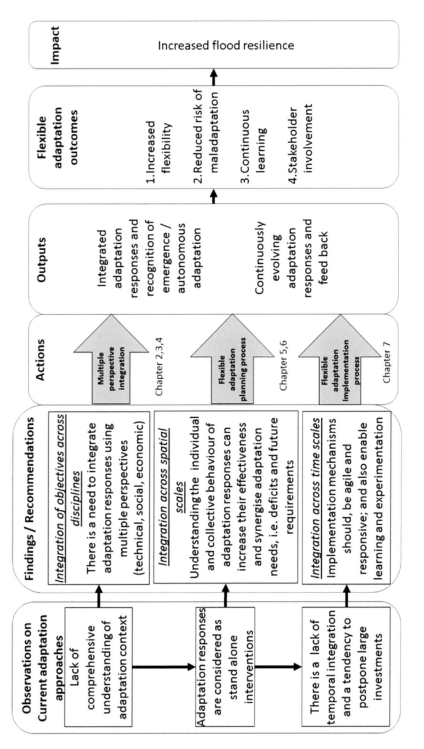

Figure 8-1 Research observations , recommendations, actions, outputs and anticipated outcomes

The local adaptation context requires an understanding of multiple adaptation needs, multiple adaptation drivers, multiple adaptation measures and their relationships among drivers and across scales from a national level to neighbourhood level or a household level. This involves understanding of the interplay of the decisions made at the national, regional or city level and how these decisions impact the adaptation at a household or neighbourhood level. Understanding the local adaptation context – by those who plan and implement adaptation measures (stakeholders) and those who support decision makers in taking decision (researchers), i.e., informed decision making – is essential, especially when there are multiple adaptation needs. Hence a framework based on a systems approach, has been presented to structure a local adaptation problems (Chapter 2). This framework is used to structure adaptation strategy responses to urban flooding in the City of Can Tho, Vietnam which is simultaneously adapting to climate change, economic change and demographic change. From using the adaptation context in Can Tho, it has been concluded that flexibility can be enhanced by ensuring compatibility between measures and taking into account the relationships between adaptation measures at various levels across spatial scales such as city level and household level (Chapter 3). For example, by including the coping capacity of households, the tipping point at which dikes no longer provide the required service can be postponed. Establishing the relationship between measures results in an increase in managerial flexibility due to an increase in the number of available adaptation measures. In the case of Can Tho, the decision makers could choose between options of elevating the dikes or subsidizing the elevation of floor-levels of vulnerable houses as the first decision point to reduce flood damage.

This research also critically looked in the role of coping capacities realized through household level measures and their emergent behavior. Presence of emergent behavior, in the coping capacities of individuals and households, in reality cannot be attributed to a single forcing such as climate and it is due to the combination and multiplicity of forcings such as climate, social, economic and political in an urban area (Jabeen et al. 2010; Thorn et al. 2015). It is hard to disaggregate the forcings that trigger autonomous adaptation (Thorn et al. 2015). For example, many reasons can be attributed to the emergence in Can Tho. One reason can be the inadequate basic services and drainage infrastructure, which causes flooding during rainfall events (SCE 2013), though the dikes are effective in preventing the fluvial floods (Chapter 3). The other reason for autonomous adaptation is the lack of trust in the local and national governments as these governing bodies are inclined towards implementation of big infrastructure adaptation measures, which overlooks ecological and social concerns (Næss et al. 2005). The current preference for the big infrastructure measures can be due to alignment of political and economic interests at national and local levels of governance. This collusion is clearly evident in case of rampant real estate development in Can Tho (Garschagen 2014). Also emergence can also be due to the cultural and economic practices rooted in the past. For example, the elevation of floor level of households in Can Tho can be attributed to the "Living with water" lifestyle that has been prevalent in Mekong Delta (Garschagen 2015; Wesselink et al. 2016).

Although the multiple adaptation problem can be structured using a multiple perspective adaptation framework as presented here, generating adaptation pathways and selecting preferred pathways remains a challenge due to a number of adaptation responses, adaptation drivers and plausible scenarios in future. Here, this challenge has been overcome by using a precedence based adaptation grammar (algorithm) for generating and evaluating adaptation pathways (Islam 2016). These pathways were generated using rules that are based on the relationship between adaptation measures and adaptation drivers (Chapter 4). Application of this methodology has been shown to lead to effective adaptation outcomes in Can Tho that are based on user defined criteria.

Mapping of relationships between adaptation measures and framing the rules for generating adaptation pathways has been possible due to an understanding of the characteristic features of the selected adaptation measures. An increased understanding from the 'singular' perspective to 'multiple perspective' of adaptation measures enables in-depth analysis of the performance and relevance of these measures in the face of multiple uncertainties. This leds to the creation of a flexible adaptation planning process (WSCapp) based on an analysis of the change of impact propagated in the urban systems due to the adaptation measures (Chapter 5). WSCapp has been developed by utilising flexible adaptation practices that are prevalent in the automobile and aerospace industries (Eckert et al. 2004; Suh et al. 2007). WSCapp has been used to identify effective adaptation responses and pathways based on change of impact propagated through an urban catchment, Elster creek in Melbourne, Australia (Chapter 6).

This thesis has contributed to an increased understanding as to how, when and where flexibility can be included in adaptation planning processes (Chapters 2 – 6). However, the work would have remained incomplete without an operational framework, that could be used to incorporate flexibility in the planning and implementation practices at a regional or city level; or by the agencies, such as Melbourne Water, who can implement, operate and maintain adaptable flood risk management systems. Flexible operational practices such as "agile approaches" (Beck et al. 2001), are common in the software development field, where urgency and uncertainties are resolved daily. This approach has been used to create an "agile urban adaptation approach" (i) for implementing and monitoring the continuously evolving adaptation measures; and (ii) to promote continuous learning and stakeholder involvement (Chapter 7). The scope for implementing the complicated agile urban adaptation approach has been explored in the complex urban adaptation context of Can Tho, which appears promising.

Although the research has analysed various adaptation perspectives and adaptation needs, it has limited its scope to urban flooding. However, it is possible to extend the findings – at a conceptual level – to other domains such as drought management, water scarcity and urban temperature variation. Future research on structuring adaptation responses or identification of flexible adaptation components can broaden the focus from flood risk management and include adaptation measures which address issues such as drought management, earthquake risks, fire hazard, health risks , increasing resilience of other critical infrastructure such as roads, power grid and tele communications. Also in this thesis, framing the adaptation problem in a local adaptation context was done only in the context of increasing or incorporating flexibility. However, the same process could also be used to assess the risk of maladaptation and help in minimising the risk of maladaptation. The research also to a lesser extent probed into the relationship between emergence (i.e., autonomous adaptation) and maladaptation.

There are contradictory views about emergence and maladaptation. Emergence is considered maladaptive and also is considered as complementary to planned adaptation (Milman and Warner 2016). Emergence is presumed to be ineffective and maladaptive as it is considered as a spontaneous activity, its effectiveness limited to a short duration and the possibility of eroding long term adaptation capacity due to frequent use of resources for short term coping (Schaer 2015; Stern 2007). However emergence is not always spontaneous and is the result of cultural and economic practices rooted in the past (Forsyth and Evans 2013). Elevation of floor levels of houses in Can Tho can be attributed to the "Living with water" life style of people in Mekong Delta (Wesselink et al. 2016). Considering coping capacity at the households due to elevation of house floor levels enhances the functional life of dikes in Can Tho, which is complementary to the planned adaptation measures (Radhakrishnan et al. 2017b).

Maladaptation of emergent responses can be attributed to the lack of basic infrastructure and services (Jabeen et al. 2010; Schaer 2015). For example, the analysis of flood water quality in Can Tho reveals that the pathogen and contaminant concentrations are as high as in the sewer (Nguyen et al. 2017), which is maladaptation due to emergence in Can Tho. An integrated assessment of adaptation responses, infrastructure and basic service at Can Tho reveals the cause of maladaptation. The poor water quality of flood water is not due to the nature of emergence at households, where as it is due to the lack of coverage or carrying capacity of drainage system in Can Tho (SCE 2013). Hence in order to prevent maladaptation the deficiency in basic services and infrastructure (Type I adaptation) have to be addressed by incorporating adaptations plans into development plans.

Hence it can be stated that the maladaptation is not due to the nature of autonomous adaptation measures. However, it is due to the (i) inadequate consideration of adaptation objectives and forcings (Thorn et al. 2015; Wamsler and Brink 2014a; Wamsler and Brink 2014b),   (ii) the lack of understanding of interface between planned and adaptation responses (Milman and Warner 2016); (iii) reductionist approach to adaptation rather than integrated approaches (Forsyth and Evans 2013); and (iv) current deficits in basic services and infrastructure (Jabeen et al. 2010; Schaer 2015). Further, public institutions cannot claim legitimacy over autonomous adaptation measures and cannot attribute maladaptation to emergence unless they address the basic infrastructure and service deficits (Mycoo 2014; Thorn et al. 2015).

Although the research has not focussed upon the aspects of changing stakeholder perceptions and governance, these aspects have been acknowledged as important components when structuring a climate adaptation problem at the planning stage and in resolving the complications during implementation. The latter implies that there is scope for further research based on the outcomes from this thesis, which could address its limitations. For example, the multiple perspective structuring of adaptation responses or the WSCapp could be extended to include the aspects of stakeholder perceptions or governance aspects for improving the adaptation responses or implementing the adaptation measures.

The thesis postulated the premise that structuring climate adaptation problem and integrating responses is an effective strategy to deal with urban flood risk in the face of uncertainty. Under the premise of adaptation as a strategy, identification of flexible adaptation components and the incorporation of flexibility in adaptation strategies are recommended as tactics. Whereas, the agile adaptation implementation practices are recommended for operationalising flexibility. Although an operational process –agile urban adaptation process – has been proposed to operationalise flexibility at a city level, the enabling factors and hindrances at a city level to achieve agile adaptation outcomes are yet to be identified.

Hence, it is concluded that an understanding of adaptation in a local context (what), relationship between the adaptation measures (how), identification of flexible adaptation measures (where) and operationalising flexibility in an agile manner (when) will increase urban flood resilience. Favourable adaptation outcomes in a changing climate context will be achieved when adaptation responses are structured using multiple perspectives and when the adaptation measures include flexibility as a key attribute.

9 References

100 Resilient Cities (2017) City Strategies. http://www.100resilientcities.org/strategies#/-_/. Accessed 27th March 2017

Adger WN, Dessai S, Goulden M, Hulme M, Lorenzoni I, Nelson D, Naess L, Wolf J, Wreford A (2009) Are there social limits to adaptation to climate change? Climatic Change 93:335-354 doi:10.1007/s10584-008-9520-z

Adger WN, Pulhin JM, Barnett J, Dabelko GD, Hovelsrud GK, Levy M, Ú. Oswald S, Vogel CH (2014) Human security. In: Field CB et al. (eds) Climate Change 2014: Impacts, Adaptation, and Vulnerability. Part A: Global and Sectoral Aspects. Contribution of Working Group II to the Fifth Assessment Report of the Intergovernmental Panel of Climate Change. Cambridge University Press, Cambridge, United Kingdom and New York, NY, USA, pp 755-791

AECOM (2012) Adapting to inundation in urbanised areas: Supporting decision makers in a changing Climate -Port Phillip Bay Coastal Adaptation Project Report. Municipal Association of Victoria, Melbourne

Aerts JCJH, Botzen WJW, Emanuel K, Lin N, de Moel H, Michel-Kerjan EO (2014) Evaluating Flood Resilience Strategies for Coastal Megacities Science 344:473-475 doi:10.1126/science.1248222

Alberti M, Marzluff JM, Shulenberger E, Bradley G, Ryan C, Zumbrunnen C (2003) Integrating Humans into Ecology: Opportunities and Challenges for Studying Urban Ecosystems BioScience 53:1169-1179 doi:10.1641/0006-3568(2003)053[1169:ihieoa]2.0.co;2

Allen C, Birge H, Bartelt-Hunt S, Bevans R, Burnett J, Cosens B, Cai X, Garmestani A, Linkov I, Scott E, Solomon M, Uden D (2016) Avoiding Decline: Fostering Resilience and Sustainability in Midsize Cities Sustainability 8:844

Anguelovski I, Chu E, Carmin J (2014) Variations in approaches to urban climate adaptation: Experiences and experimentation from the global South Global Environmental Change 27:156-167 doi:http://dx.doi.org/10.1016/j.gloenvcha.2014.05.010

Anvarifara F, Zevenbergen C, Thissen W, Islam T (2016) Understanding flexibility for multifunctional flood defences: a conceptual framework Journal of Water and Climate Change doi:10.2166/wcc.2016.064

Apel H, Martínez Trepat O, Hung NN, Chinh DT, Merz B, Dung NV (2016) Combined fluvial and pluvial urban flood hazard analysis: concept development and application to Can Tho city, Mekong Delta, Vietnam Natural Hazards and Earth System Sciences 16:941

Ashley MR, Digman CJ, Horton B Demonstrating and monetizing the multiple benefits from using sustainable drainage. In: World Water Congress, Brisbane, Australia, 2016. IWA,

Ashley RM, Digman CJ, Horton B, Gersonius B, Smith B, Shaffer P, Baylis A (In Press) Evaluating the longer term benefits of sustainable drainage Proceedings of the Institution of Civil Engineers - Water Management 0:1-10 doi:10.1680/jwama.16.00118

Ashley R (2012) Interactions with flood affected people of Brisbane 2011 floods. Richard Ashley, Brisbane

Ashley R, Blanksby J, Chapman J, Zhou J (2007) Towards integrated approaches to reduce flood risk in urban areas Advances in Urban Flood Management: 415-432

Ashley R, Gersonius B, Blanksby J, Stam J-M, Doncaster S (2013a) Water sensitive urban design for the flood resilient city. Paper presented at the 8th International Conference on planning & technologies for sustainable urban water management, Lyon, France, June 23-27, 2013

Ashley R, Lundy L, Ward S, Shaffer P, Walker L, Morgan C, Saul A, Wong T, Moore S (2013b) Water-sensitive urban design: opportunities for the UK Proceedings of the Institution of Civil Engineers - Municipal Engineer 166:65-76 doi:10.1680/muen.12.00046

Ashley R, Walker L, D'Arcy B, Wilson S, Illman S, Shaffer P, Woods-Ballard B, Chatfield P (2015) UK sustainable drainage systems: past, present and future Proceedings of the Institution of Civil Engineers - Civil Engineering 168:125-130 doi:10.1680/cien.15.00011

Ashley RM, Blanskby J, Newman R, Gersonius B, Poole A, Lindley G, Smith S, Ogden S, Nowell R (2012) Learning and Action Alliances to build capacity for flood resilience Journal of Flood Risk Management 5:14-22 doi:10.1111/j.1753-318X.2011.01108.x

Balmforth D, Digman C, Kellagher R, Butler D (2006) CIRIA C635: Designing for exceedance in urban drainage – good practice. CIRIA, London

Barnett J, O'Neill S (2010) Maladaptation Global Environmental Change 20:211-213

Beck K, Beedle M, van Bennekum A, Cockburn A, Cunningham W, Fowler M, Grenning J, Highsmith J, Hunt A, Jeffries R, Kern J, Marick B, Martin R, Mallor S, Shwaber K, Sutherland J (2001) The Agile Manifesto. Agile manifesto.org. http://agilemanifesto.org/. Accessed 06/10/2016 2016

Bek M, Bugra A, Hjalmarsson J, Lista A (2013) Future availability of flood insurance in UK: A report on legal aspects of the solutions adopted in Australia, Iceland, the Netherlands, New Zealand and Turkey, with conclusions

Bernardes ES, Hanna MD (2009) A theoretical review of flexibility, agility and responsiveness in the operations management literature: Toward a conceptual definition of customer responsiveness International Journal of Operations & Production Management 29:30-53 doi: 10.1108/01443570910925352

Bilalić M, McLeod P, Gobet F (2008) Why good thoughts block better ones: The mechanism of the pernicious Einstellung (set) effect Cognition 108:652-661 doi:http://dx.doi.org/10.1016/j.cognition.2008.05.005

Birkmann J, Garschagen M, Kraas F, Quang N (2010) Adaptive urban governance: new challenges for the second generation of urban adaptation strategies to climate change Sustainability Science 5:185-206

Birkmann J, Garschagen M, Van Tuan V, Binh N (2012) Vulnerability, Coping and Adaptation to Water Related Hazards in the Vietnamese Mekong Delta. In: Renaud FG, Kuenzer C (eds) The Mekong Delta System. Springer Environmental Science and Engineering. Springer Netherlands, pp 245-289. doi:10.1007/978-94-007-3962-8_10

Bowen A, Cochrane S, Fankhauser S (2012) Climate change, adaptation and economic growth Climatic Change 113:95-106 doi:10.1007/s10584-011-0346-8

Brisley R, Wylde R, Lamb R, Cooper J, Sayers P, Hall J (2016) Techniques for valuing adaptive capacity in flood risk management Proceedings of the Institution of Civil Engineers - Water Management 169:75-84 doi:10.1680/jwama.14.00070

Brown R, Keath N, Wong T (2009) Urban water management in cities: historical, current and future regimes Water Science & Technology—WST 59:847–855

Bryant BP, Lempert RJ (2010) Thinking inside the box: A participatory, computer-assisted approach to scenario discovery Technological Forecasting and Social Change 77:34-49 doi:http://dx.doi.org/10.1016/j.techfore.2009.08.002

Burkett VR, Suarez AG, Bindi M, Conde C, Mukerji R, Prather MJ, Clair ALS, Yohe GW (2014) Point of departure. In: Field CB et al. (eds) Climate Change 2014: Impacts, Adaptation, and Vulnerability. Part A: Global and Sectoral Aspects. Contribution of Working Group II to the Fifth Assessment Report of the Intergovernmental Panel of Climate Change. Cambridge University Press, Cambridge, United Kingdom and New York, NY, USA, pp 169-194

Burton I Climate Change and Adaptation Deficit. In: Adam Fenech RBR, Don MacIver and Heather Auld (ed) International Conference on Adaptation Science, Management and Policy Options, Lijiang, Yunnan, China, May 17-19, 2004 2004. AIRG, Meteorogical Service of Canada, Environment Canada, pp 25-33

Buuren A, Driessen P, Teisman G, Rijswick M (2013) Toward legitimate governance strategies for climate adaptation in the Netherlands: combining insights from a legal, planning, and network perspective Regional Environmental Change 14:1021-1033 doi:10.1007/s10113-013-0448-0

Buurman J, Babovic V (2016) Adaptation Pathways and Real Options Analysis: An approach to deep uncertainty in climate change adaptation policies Policy and Society 35:137-150 doi:http://dx.doi.org/10.1016/j.polsoc.2016.05.002

Calabrese C, Baresi L (2017) Outdoor Augmented Reality for Urban Design and Simulation. In: Piga BEA, Salerno R (eds) Urban Design and Representation: A Multidisciplinary and Multisensory Approach. Springer International Publishing, Cham, pp 181-190. doi:10.1007/978-3-319-51804-6_14

Carus P (1892) WHAT DOES ANSCHAUUNG MEAN? The Monist 2:527-532

Cettner A, Ashley R, Hedström A, Viklander M (2014a) Assessing receptivity for change in urban stormwater management and contexts for action Journal of Environmental Management 146:29-41 doi:http://dx.doi.org/10.1016/j.jenvman.2014.07.024

Cettner A, Ashley R, Hedström A, Viklander M (2014b) Sustainable development and urban stormwater practice Urban Water Journal 11:185-197 doi:10.1080/1573062x.2013.768683

Chang M Agile and Crystal Clear with Library IT Innovations. In: VALA2010 Conference, 2010.

CHI (2017) PCSWMM Software. Computational Hydralics International, Ontario, Canada

Chinh D, Gain A, Dung N, Haase D, Kreibich H (2016a) Multi-Variate Analyses of Flood Loss in Can Tho City, Mekong Delta Water 8:6

Chinh DT, Dung NV, Kreibich H, Bubeck P (2016b) The 2011 flood event in the Mekong Delta: preparedness, response, damage and recovery of private households and small businesses Disasters:n/a-n/a doi:10.1111/disa.12171

Chu E, Anguelovski I, Carmin J (2015) Inclusive approaches to urban climate adaptation planning and implementation in the Global South Climate Policy:1-21 doi:10.1080/14693062.2015.1019822

Chu E, Anguelovski I, Roberts D (2017) Climate adaptation as strategic urbanism: assessing opportunities and uncertainties for equity and inclusive development in cities Cities 60, Part A:378-387 doi:http://dx.doi.org/10.1016/j.cities.2016.10.016

Cilliers P (2001) BOUNDARIES, HIERARCHIES AND NETWORKS IN COMPLEX SYSTEMS International Journal of Innovation Management 05:135-147 doi: 10.1142/s1363919601000312

City of Melbourne (2016) Resilient Melbourne. City of Melbourne, Melbourne

Clemens M, Rijke J, Pathirana A, Evers J, Hong Quan N (2015) Social learning for adaptation to climate change in developing countries: insights from Vietnam Journal of Water and Climate Change doi:10.2166/wcc.2015.004

Clemens M, Rijke J, Pathirana A, Quan NH (2014) Effective Stakeholder Analysis for Urban Flood Resilience in Vietnam Using Design Proposition. Paper presented at the 13th International Conference on Urban Drainage, Sarawak, Malaysia, 7 – 12 September 2014

Commonwealth of Australia G (2010) Adapting to Climate Change in Australia -An Australian Government Position Paper.

CoPP (2016) Amendment C111 - SBO review. City of Port Phillip,Melbourne, Australia. http://www.portphillip.vic.gov.au/amendment-c111.htm#What is the SBO? Accessed 15 Dec 2016 2016

CRIDA (In press) Water Resources Planning & Design for an Uncertain Future. International Center for Integrated Water Resources Management, ICIWaRM Press, Alexandria, Virginia, USA

CSIRO (2015) Climate Change in Australia Information for Australia's Natural Resource Management Regions: Technical Report. CSIRO and Bureau of Meteorology, Australia,

Davidsen S, Löwe R, Thrysøe C, Arnbjerg-Nielsen K (2017) Simplification of one-dimensional hydraulic networks by automated processes evaluated on 1D/2D deterministic flood models Journal of Hydroinformatics Accepted

De Neufville R, Scholtes S (2011) Flexibility in engineering design. The MIT Press,

De Sherbinin A, Schiller A, Pulsipher A (2007) The vulnerability of global cities to climate hazards Environment and Urbanization 19:39-64

defra (2009) Accounting for Effects of Climate Change - Supplementary Green Book Guidance HM Treasury and DEFRA, London

Deltacommissaris (2014) Delta Prgramme 2015 : Working on the Dutch Delta in the 21st century : A new phase in the battle against the water. The Ministry of Infrastructure and Environment, The Ministry of Economic Afffairs , The Hague

Dessai S, Sluijs JP (2007) Uncertainty and climate change adaptation: A scoping study vol 2007. Copernicus Institute for Sustainable Development and Innovation, Department of Science Technology and Society Utrecht, the Netherlands,

Dewulf A (2013) Contrasting frames in policy debates on climate change adaptation Wiley Interdisciplinary Reviews: Climate Change 4:321-330 doi:10.1002/wcc.227

Dewulf A, Craps M, Bouwen R, Taillieu T, Pahl-Wostl C (2005) Integrated management of natural resources: dealing with ambiguous issues, multiple actors and diverging frames Water Science & Technology 52:115-124

Digman CJ, Ashley R (2014) Managing urban flooding from heavy rainfall – encouraging the uptake of designing for exceedance – Recommendations and Summary vol CIRIA RP991. Construction Industry Research and Information Association, London

Dittrich R, Wreford A, Moran D (2016) A survey of decision-making approaches for climate change adaptation: Are robust methods the way forward? Ecological Economics 122:79-89 doi:http://dx.doi.org/10.1016/j.ecolecon.2015.12.006

Dixit AK, Pindyck RS (1994) Investment under Uncertainty. Princeton University Press, New Jersey

Dunn G, Brown R, Bos JJ, Bakker K (2016) Standing on the Shoulders of the Gaints: Understanding changes in urban water practices through the lens of complexity sciences Urban water doi:10.1080/1573062X.2016.1241284

Dupuis J, Knoepfel P (2013) The Adaptation Policy Paradox: the Implementation Deficit of Policies Framed as Climate Change Adaptation Ecology and society 18 doi: 10.5751/es-05965-180431

DWF (2011) Survey on Perception of risk in Can Tho City. Development Workshop France, Lauzerte, France

Ebi KL, Semenza JC, Rocklöv J (2016) Current medical research funding and frameworks are insufficient to address the health risks of global environmental change Environmental Health 15:108 doi:10.1186/s12940-016-0183-3

Eckert C, Clarkson PJ, Zanker W (2004) Change and customisation in complex engineering domains Research in Engineering Design 15:1-21 doi:10.1007/s00163-003-0031-7

Edelenbos J (2005) Institutional Implications of Interactive Governance: Insights from Dutch Practice Governance 18:111-134 doi:10.1111/j.1468-0491.2004.00268.x

EEA (2016) Urban adaptation to climate change in Europe: Transforming Cities in a changing climate. European Environment Agency, Copenhagen

Ellen GJ, van Leeuwen C, Kuindersma W, Breman B, van Lamoen F (2014) 7 Adaptive governance in practice Action Research for Climate Change Adaptation: Developing and Applying Knowledge for Governance:112

Eriksen SH, Nightingale AJ, Eakin H (2015) Reframing adaptation: The political nature of climate change adaptation Global Environmental Change 35:523-533 doi:http://dx.doi.org/10.1016/j.gloenvcha.2015.09.014

EU (2007) directive 2007/60/EC of the European Parliament and of the council of 23 October 2007 on the assessment and management of flood risks. European Union,

Felgenhauer T, Webster M (2013) Multiple adaptation types with mitigation: A framework for policy analysis Global Environmental Change 23:1556-1565 doi:http://dx.doi.org/10.1016/j.gloenvcha.2013.09.018

Ferguson BC, Brown RR, Frantzeskaki N, de Haan FJ, Deletic A (2013a) The enabling institutional context for integrated water management: Lessons from Melbourne Water research 47:7300-7314 doi:http://dx.doi.org/10.1016/j.watres.2013.09.045

Ferguson BC, Frantzeskaki N, Brown RR (2013b) A strategic program for transitioning to a Water Sensitive City Landscape and Urban Planning 117:32-45 doi:http://dx.doi.org/10.1016/j.landurbplan.2013.04.016

Fletcher TD, Shuster W, Hunt WF, Ashley R, Butler D, Arthur S, Trowsdale S, Barraud S, Semadeni-Davies A, Bertrand-Krajewski J-L, Mikkelsen PS, Rivard G, Uhl M, Dagenais D, Viklander M (2015) SUDS, LID, BMPs, WSUD and more – The evolution and application of terminology surrounding urban drainage Urban Water Journal 12:525-542 doi:10.1080/1573062x.2014.916314

Floyd RW (1963) Syntactic Analysis and Operator Precedence J ACM 10:316-333 doi:10.1145/321172.321179

Forsyth T, Evans N (2013) What is Autonomous Adaption? Resource Scarcity and Smallholder Agency in Thailand World Development 43:56-66 doi:http://doi.org/10.1016/j.worlddev.2012.11.010

Fowler M, Highsmith J (2001) The agile manifesto Software Development 9:28-35

Fratini CF, Geldof GD, Kluck J, Mikkelsen PS (2012) Three Points Approach (3PA) for urban flood risk management: A tool to support climate change adaptation through transdisciplinarity and multifunctionality Urban Water Journal 9:317-331 doi:10.1080/1573062x.2012.668913

Fricke E, Schulz AP (2005) Design for changeability (DfC): Principles to enable changes in systems throughout their entire lifecycle Systems Engineering 8:no-no doi:10.1002/sys.20039

Füssel H-M (2007) Adaptation planning for climate change: concepts, assessment approaches, and key lessons Sustainability Science 2:265-275

Gandhi G (2015) The Thriukkural. Aleph Book Company, New Delhi, India

Garschagen M (2014) Risky change? Vulnerability and adaptation between climate change and transformation dynamics in Can Tho City, Vietnam vol 15. Megacities and Global Change. Steiner, Stuttgart

Garschagen M (2015) Risky Change? Vietnam's Urban Flood Risk Governance between Climate Dynamics and Transformation Pacific Affairs 88:599-621 doi:10.5509/2015883599

Garschagen M, Romero-Lankao P (2013) Exploring the relationships between urbanization trends and climate change vulnerability Climatic Change 133:37-52 doi:10.1007/s10584-013-0812-6

Geldof GD (1995) Adaptive water management: Integrated water management on the edge of chaos Water Science and Technology 32:7-13 doi:http://dx.doi.org/10.1016/0273-1223(95)00532-R

Gersonius B, Ashley R, Pathirana A, Zevenbergen C (2012a) Adaptation of flood risk infrastructure to climate resilience Proceedings of the Institution of Civil Engineers - Civil Engineering 165:40-45 doi:10.1680/cien.11.00053

Gersonius B, Ashley R, Pathirana A, Zevenbergen C (2013) Climate change uncertainty: building flexibility into water and flood risk infrastructure Climatic Change 116:411-423 doi:10.1007/s10584-012-0494-5

Gersonius B, Ashley R, Salinas Rodriguez CNA, Rijke J, Radhakrishnan M, Zevenbergen C (2016) Flood Resilience in Water Sensitive Cities. Cooperative Research Centre for Water Sensitive Cities, Clayton, Melbourne, Australia

Gersonius B, Morselt T, van Nieuwenhuijzen L, Ashley R, Zevenbergen C (2012b) How the Failure to Account for Flexibility in the Economic Analysis of Flood Risk and Coastal

Management Strategies Can Result in Maladaptive Decisions Journal of Waterway, Port, Coastal, and Ocean Engineering 138:386-393 doi:doi:10.1061/(ASCE)WW.1943-5460.0000142

Gersonius B, Nasruddin F, Ashley R, Jeuken A, Pathirana A, Zevenbergen C (2012c) Developing the evidence base for mainstreaming adaptation of stormwater systems to climate change Water research 46:6824-6835 doi:http://dx.doi.org/10.1016/j.watres.2012.03.060

GHD (2014) Elwood Canal Catchment Flood Mitigation - Stage 2 : Mitiagion option assessment - Final report Melbourne Water Melbourne

Groves DG, Lempert RJ (2007) A new analytic method for finding policy-relevant scenarios Global Environmental Change 17:73-85 doi:http://dx.doi.org/10.1016/j.gloenvcha.2006.11.006

Gunn A, Rogers BC (2015) Towards a water sensitive Elwood: a community vision and transition pathways. CRC for Water Sensitive Cities, Melbourne

Haasnoot M, Kwakkel JH, Walker WE, ter Maat J (2013) Dynamic adaptive policy pathways: A method for crafting robust decisions for a deeply uncertain world Global Environmental Change 23:485-498 doi:http://dx.doi.org/10.1016/j.gloenvcha.2012.12.006

Haasnoot M, Middelkoop H, Offermans A, Beek E, Deursen WP (2012a) Exploring Pathways for Sustainable Water Management in River Deltas in a Changing Environment Clim Change 115 doi:10.1007/s10584-012-0444-2

Haasnoot M, Middelkoop H, Offermans A, Beek E, Deursen WPAv (2012b) Exploring pathways for sustainable water management in river deltas in a changing environment Climatic Change 115:795-819 doi:10.1007/s10584-012-0444-2

Haasnoot M, Middelkoop H, van Beek E, van Deursen WPA (2011) A method to develop sustainable water management strategies for an uncertain future Sustainable Development 19:369-381 doi:10.1002/sd.438

Haasnoot M, Van Deursen W (2015) Pathways Generator. Deltares and Carthago Consultancy. https://publicwiki.deltares.nl/display/AP/Adaptation+Pathways. Accessed 21 Sep 2016 2016

Haasnoot M, van Deursen WPA, Guillaume JHA, Kwakkel JH, van Beek E, Middelkoop H (2014) Fit for purpose? Building and evaluating a fast, integrated model for exploring water policy pathways Environmental Modelling & Software 60:99-120 doi:http://dx.doi.org/10.1016/j.envsoft.2014.05.020

Hall JW, Lempert RJ, Keller K, Hackbarth A, Mijere C, McInerney DJ (2012) Robust Climate Policies Under Uncertainty

Hallegatte S, Henriet F, Corfee-Morlot J (2011) The economics of climate change impacts and policy benefits at city scale: a conceptual framework Climatic Change 104:51-87 doi:10.1007/s10584-010-9976-5

Haque AN, Dodman D, Hossain MM (2014) Individual, communal and institutional responses to climate change by low-income households in Khulna, Bangladesh Environment and Urbanization 26:112-129 doi:doi:10.1177/0956247813518681

Hassan FA (1997) Nile Floods and Political Disorder in Early Egypt. In: Dalfes HN, Kukla G, Weiss H (eds) Third Millennium BC Climate Change and Old World Collapse. Springer Berlin Heidelberg, Berlin, Heidelberg, pp 1-23. doi:10.1007/978-3-642-60616-8_1

Hinkel J, Bots PWG, Schlüter M (2014) Enhancing the Ostrom social-ecological system framework through formalization Ecology and society 19 doi:10.5751/es-06475-190351

HM Government (2016) National Flood Resilience Review. Crown copyright 2016,

Holland HJ (1992) Complex Adaptive Systems Daedalus 121

Holweg M (2005) The three dimensions of responsiveness International Journal of Operations & Production Management 25:603-622 doi:doi:10.1108/01443570510605063

Horton B, Digman CJ, Ashley RM, Gill E (2016) BeST (Benefits of SuDS Tool) W045c BeST - Technical Guidance. Release Version 3, vol RP993. CIRIA, Griffin Court, 15 Long Lane, London, EC1A 9PN, UK

Howe C, Mitchell C (2011) Water Sensitive Cities Water Intelligence Online 10 doi:10.2166/9781780400921

Huong H, Pathirana A (2013) Urbanization and climate change impacts on future urban flood risk in Can Tho city, Vietnam Hydrol Earth Syst Sci Discuss 17:379-394 doi:10.5194/hess-17-379-2013

Hurley L, Ashley R, Mounce S (2008) Addressing practical problems in sustainability assessment frameworks Proceedings of the Institution of Civil Engineers - Engineering Sustainability 161:23-30 doi:doi:10.1680/ensu.2008.161.1.23

IPCC (2007) Climate Change 2007: Impacts, Adaptation and Vulnerability Contribution of Working Group II to the Fourth Assessment Report of the Intergovernmental Panel on Climate Change. Cambridge University Press, Cambridge ,UK

IPCC (2013) Working Group I Contribution to the IPCC Fifth Assessment Report, Climate Change 2013: The Physical Science Basis, Summary for Policymakers IPCC, Geneva, Switzerland

IPCC (2014a) Climate Change 2014: Impacts, Adaptation, and Vulnerability. Part A: Global and Sectoral Aspects. Contribution of Working Group II to the Fifth Assessment Report of the Intergovernmental Panel on Climate Change [Field, C.B., V.R. Barros, D.J. Dokken, K.J. Mach, M.D. Mastrandrea, T.E. Bilir, M. Chatterjee, K.L. Ebi, Y.O. Estrada, R.C. Genova, B. Girma, E.S. Kissel, A.N. Levy, S. MacCracken, P.R. Mastrandrea, and L.L. White (eds.)]. Cambridge University Press, Cambridge, United Kingdom and New York, NY, USA

IPCC (2014b) Climate Change 2014: Impacts, Adaptation, and Vulnerability. Part B: Regional Aspects. Contribution of Working Group II to the Fifth Assessment Report of the Intergovernmental Panel on Climate Change [Barros, V.R., C.B. Field, D.J. Dokken, M.D. Mastrandrea, K.J. Mach, T.E. Bilir, M. Chatterjee, K.L. Ebi, Y.O. Estrada, R.C. Genova, B. Girma, E.S. Kissel, A.N. Levy, S. MacCracken, P.R. Mastrandrea, and L.L. White (eds.)]. Cambridge University Press, Cambridge, United Kingdom and New York, NY, USA

IPCC (2014c) Climate Change 2014: Impacts, Adaptation, and Vulnerability. Summaries, Frequently Asked Questions, and Cross-Chapter Boxes. A Contribution of Working Group II to the Fifth Assessment Report of the Intergovernmental Panel on Climate Change [Field, C.B., V.R. Barros, D.J. Dokken, K.J. Mach, M.D. Mastrandrea, T.E. Bilir, M. Chatterjee, K.L. Ebi, Y.O. Estrada, R.C. Genova, B. Girma, E.S. Kissel, A.N. Levy, S. MacCracken,P.R. Mastrandrea, and L.L. White (eds.)]:190

IPCC (2014d) Summary for Policymakers. In: Field CB et al. (eds) Climate Change 2014: Impacts, Adaptation, and Vulnerability. Part A: Global and Sectoral Aspects. Contribution of Working Group II to the Fifth Assessment Report of the Intergovernmental Panel on

Climate Change. Cambridge University Press, Cambridge, United Kingdom, and New York, NY, USA, pp 1-32

Islam T (2016) Modelling and evaluation of modular systems using context dependent grammars Paper presented at the CESUN 2016, 5th International Engineering Systems Symposium, Washington DC, June 27-29 2016

Jabareen Y (2013) Planning the resilient city: Concepts and strategies for coping with climate change and environmental risk Cities 31:220-229 doi:http://dx.doi.org/10.1016/j.cities.2012.05.004

Jabeen H, Johnson C, Allen A (2010) Built-in resilience: learning from grassroots coping strategies for climate variability Environment and Urbanization 22:415-431 doi:doi:10.1177/0956247810379937

James W, Huber W, Pitt R, Dickinson R, Rosener L, Aldrich J, James W (2002) SWMM4 User's Manual (User's Guide to the EPA Stormwater Management Model and to PCSWMM). University of Guelph and CHAI,

Jiang L, O'Neill BC (2015) Global urbanization projections for the Shared Socioeconomic Pathways Global Environmental Change doi:http://dx.doi.org/10.1016/j.gloenvcha.2015.03.008

Jonas E, Traut-Mattausch E, Frey D, Greenberg J (2008) The path or the goal? Decision vs. information focus in biased information seeking after preliminary decisions Journal of Experimental Social Psychology 44:1180-1186 doi:http://dx.doi.org/10.1016/j.jesp.2008.02.009

Jonkman SN, Hillen MM, Nicholls RJ, Kanning W, van Ledden M (2013) Costs of Adapting Coastal Defences to Sea-Level Rise— New Estimates and Their Implications Journal of Coastal Research:1212-1226 doi:10.2112/jcoastres-d-12-00230.1

Kasprzyk JR, Reed PM, Hadka DM (2016) Battling Arrow's Paradox to Discover Robust Water Management Alternatives Journal of Water Resources Planning and Management 142:04015053 doi:doi:10.1061/(ASCE)WR.1943-5452.0000572

Kettunen P (2009) Adopting key lessons from agile manufacturing to agile software product development—A comparative study Technovation 29:408-422 doi:http://dx.doi.org/10.1016/j.technovation.2008.10.003

Kind JM (2014) Economically efficient flood protection standards for the Netherlands Journal of Flood Risk Management 7:103-117 doi:10.1111/jfr3.12026

Klein RJT, Adams KM, Dzebo A, Davis M, Siebert CK (2017) Advancing climate adaptation practices and solutions: Emerging research priorities vol 2017. Stockholm Enviroment Institute, Stockholm

Kleinert S, Horton R (2016) Urban design: an important future force for health and wellbeing The Lancet 388:2848-2850 doi:10.1016/s0140-6736(16)31578-1

Klijn F, Kreibich H, de Moel H, Penning-Rowsell E (2015) Adaptive flood risk management planning based on a comprehensive flood risk conceptualisation Mitig Adapt Strateg Glob Change:1-20 doi:10.1007/s11027-015-9638-z

Koste LL, Malhotra MK (1999) A theoretical framework for analyzing the dimensions of manufacturing flexibility Journal of Operations Management 18:75-93 doi:http://dx.doi.org/10.1016/S0272-6963(99)00010-8

Kreibich H, Bubeck P, Van Vliet M, De Moel H (2015) A review of damage-reducing measures to manage fluvial flood risks in a changing climate Mitig Adapt Strateg Glob Change:1-23 doi:10.1007/s11027-014-9629-5

Kwadijk JCJ, Haasnoot M, Mulder JPM, Hoogvliet MMC, Jeuken ABM, van der Krogt RAA, van Oostrom NGC, Schelfhout HA, van Velzen EH, van Waveren H, de Wit MJM (2010) Using adaptation tipping points to prepare for climate change and sea level rise: a case study in the Netherlands Wiley Interdisciplinary Reviews: Climate Change 1:729-740 doi:10.1002/wcc.64

Kwakkel JH, Haasnoot M, Walker WE (2015) Developing dynamic adaptive policy pathways: a computer-assisted approach for developing adaptive strategies for a deeply uncertain world Climatic Change 132:373-386 doi:10.1007/s10584-014-1210-4

Kwakkel JH, Haasnoot M, Walker WE (2016) Comparing Robust Decision-Making and Dynamic Adaptive Policy Pathways for model-based decision support under deep uncertainty Environmental Modelling & Software 86:168-183 doi:http://dx.doi.org/10.1016/j.envsoft.2016.09.017

Kwakkel JH, Walker WE, Marchau V (2010) Adaptive airport strategic planning European Journal of Transport and Infrastructure Research (EJTIR), 10 (3), 2010

Leffingwell D (2010) Agile software requirements: lean requirements practices for teams, programs, and the enterprise. Addison-Wesley Professional,

Leimbach M, Kriegler E, Roming N, Schwanitz J (2015) Future growth patterns of world regions – A GDP scenario approach Global Environmental Change doi:http://dx.doi.org/10.1016/j.gloenvcha.2015.02.005

Lempert RJ (2003) Shaping the next one hundred years: new methods for quantitative, long-term policy analysis. Rand Corporation,

Löwe R, Urich C, Kulahci M, Radhakrishnan M, Deletic A, Arnbjerg-Nielsen K Setup for Scenario-free Modelling of Urban Flood Risk in Non-stationary Climate and Urban Development Conditions. In: 14th IWA/IAHR International Conference on Urban Drainage, Prague, Czech Republic, 10-15 September 2017 2017. IWA Publishing,

Löwe R, Urich C, Sto Domingo N, Wong V, Mark O, Deletic A, Arnbjerg-Nielsen K Flood risk assessment as an integral part of urban planning. In: 2nd Water Sensitive Cities Conference, 2015.

Lu N, Korman T (2010) Implementation of Building Information Modeling (BIM) in Modular Construction: Benefits and Challenges. Paper presented at the Construction Research Congress 2010, Banff, Alberta, Canada

Magnan AK, Schipper ELF, Burkett M, Bharwani S, Burton I, Eriksen S, Gemenne F, Schaar J, Ziervogel G (2016) Addressing the risk of maladaptation to climate change Wiley Interdisciplinary Reviews: Climate Change 7:646-665 doi:10.1002/wcc.409

Maier HR, Guillaume JHA, van Delden H, Riddell GA, Haasnoot M, Kwakkel JH (2016) An uncertain future, deep uncertainty, scenarios, robustness and adaptation: How do they fit together? Environmental Modelling & Software 81:154-164 doi:http://dx.doi.org/10.1016/j.envsoft.2016.03.014

Malekpour S, Brown RR, de Haan FJ (2015) Strategic planning of urban infrastructure for environmental sustainability: Understanding the past to intervene for the future Cities 46:67-75 doi:http://dx.doi.org/10.1016/j.cities.2015.05.003

Manocha N, Babovic V (2016) Planning Flood Risk Infrastructure Development under Climate Change Uncertainty Procedia Engineering 154:1406-1413 doi:http://dx.doi.org/10.1016/j.proeng.2016.07.511

Mathijs van V, Jasper E, Arwin van B (eds) (2014) Action Research for Climate Change Adaptation. Routledge. doi:doi:10.4324/9781315780368

Matteo G, Daniela A, Andrea C, Phuong Nam V, Rodolfo S-S (2016) Large storage operations under climate change: expanding uncertainties and evolving tradeoffs Environmental Research Letters 11:035009

Maurer M (2013) Full costs,(dis-) economies of scale and the price of uncertainty. IWA Publishing, London,

McGaughey RE (1999) Internet technology: contributing to agility in the twenty-first century International Journal of Agile Management Systems 1:7-13 doi:doi:10.1108/14654659910266655

MDP (2013) Mekong Delta Plan - Long-term vision and strategy for a safe, prosperous and sustainable delta. Ministry of Natural Rescources and Environment - Vietnam, Ministry of Agriculture and Rural Development - Vietnam, Kingdom of Netherlands, Consortium of Royal HaskoningDHV, Wageningen University and Research Centre, Deltares, Rebel, Water.nl, Amersfoot, Nertherlands

Melbourne Water (2015) Flood Management Strategy Port Phillip and Westernport. Melbourne Water, Melbourne

Meredith S, Francis D (2000) Journey towards agility: the agile wheel explored The TQM Magazine 12:137-143 doi:doi:10.1108/09544780010318398

Milly PCD, Betancourt J, Falkenmark M, Hirsch RM, Kundzewicz ZW, Lettenmaier DP, Stouffer RJ (2008) Stationarity Is Dead: Whither Water Management? Science 319:573-574 doi:10.1126/science.1151915

Milman A, Warner BP (2016) The interfaces of public and private adaptation: Lessons from flooding in the Deerfield River Watershed Global Environmental Change 36:46-55 doi:http://doi.org/10.1016/j.gloenvcha.2015.11.007

Moench M, Tyler S, Lage J (2011) Catalyzing Urban Climate Resilience: Applying Resilience Concepts to Planning Practice in the ACCCRN Program (2009-2011). Institute for Social and Environmental Transition, International,

Moglia M, Tjandraatmadja G, Delbridge N, Gulizia E, Sharma A, Butler R, Gan K (2014) Survey of savings and conditions of rainwater tanks Melbourne, Smart Water Fund and CSIRO, Australia

Mycoo MA (2014) Autonomous household responses and urban governance capacity building for climate change adaptation: Georgetown, Guyana Urban Climate 9:134-154 doi:http://doi.org/10.1016/j.uclim.2014.07.009

Myers S (1984) Finance Theory and Financial Strategy Interfaces 14:126-137 doi:10.1287/inte.14.1.126

Næss LO, Bang G, Eriksen S, Vevatne J (2005) Institutional adaptation to climate change: Flood responses at the municipal level in Norway Global Environmental Change 15:125-138 doi:http://doi.org/10.1016/j.gloenvcha.2004.10.003

Newman R, Ashley R, Molyneux-Hodgson S, Cashman A (2011) Managing water as a socio-technical system: the shift from 'experts' to 'alliances' Proceedings of the ICE-Engineering Sustainability 164:95-102

Nguyen HQ, Radhakrishnan M, Huynh TTN, Baino-Salingay ML, Ho LP, Steen PVd, Pathirana A (2017) Water Quality Dynamics of Urban Water Bodies during Flooding in Can Tho City, Vietnam Water 9:260

Nilubon P, Veerbeek W, Zevenbergen C (2016) Amphibious Architecture and Design: A Catalyst of Opportunistic Adaptation? – Case Study Bangkok Procedia - Social and Behavioral Sciences 216:470-480 doi:http://dx.doi.org/10.1016/j.sbspro.2015.12.063

O'Brien K (2015) Political agency: The key to tackling climate change Science 350:1170-1171 doi:10.1126/science.aad0267

O'Neill BC, Kriegler E, Ebi KL, Kemp-Benedict E, Riahi K, Rothman DS, van Ruijven BJ, van Vuuren DP, Birkmann J, Kok K, Levy M, Solecki W (2015) The roads ahead: Narratives for shared socioeconomic pathways describing world futures in the 21st century Global Environmental Change doi:http://dx.doi.org/10.1016/j.gloenvcha.2015.01.004

Offermans A, Haasnoot M, Valkering P (2011) A method to explore social response for sustainable water management strategies under changing conditions Sustainable Development 19:312-324 doi:10.1002/sd.439

Olesen L, Löwe R, Arnbjerg-Nielsen K (2017) Flood damage assessment: Literature review and recommended procedure. Cooperative Research Centre for Water Sensitive Cities, Melbourne, Australia

Oppermann R (1994) Adaptively supported adaptability International Journal of Human-Computer Studies 40:455-472

Ostrom E (2009) A General Framework for Analyzing Sustainability of Social-Ecological Systems Science 325:419-422 doi:10.1126/science.1172133

Pathirana A, Radhakrishnan M, Ashley R, Quan NH, Zevenbergen C (2017-a) Managing urban water systems with significant adaptation deficits – unified framework for secondary cities : Part II – the practice

Pathirana A, Radhakrishnan M, Quan NH, Zevenbergen C (2017-b) Managing urban water systems with significant adaptation deficits – unified framework for secondary cities : part I - conceptual framework Climatic Change:1-14 doi:10.1007/s10584-017-1953-9

Pearsall J (1998) The new Oxford dictionary of English. First edn. Clarendon Press,

Peters GB (2005) The Problem of Policy Problems Journal of Comparative Policy Analysis: Research and Practice 7:349-370 doi:10.1080/13876980500319204

Pham CH, Ehlers E, Subramanian SV (2009) Dyke system planing: Theory and practice in Can Tho City, Vietnam. ZEF Working Paper Series,

Phi HL, Hermans LM, Douven WJAM, Van Halsema GE, Khan MF (2015) A framework to assess plan implementation maturity with an application to flood management in Vietnam Water International 40:984-1003 doi:10.1080/02508060.2015.1101528

PM (2013) Decision 567, 568 dated 14 September 2013 of the Prime Minister (PM) approving Socio Economic Development Plan of Can Tho city till 2020 and Vision 2030. Prime Ministers Office, Ha Noi

Pope GU, Drew WH, Lazarus J, Ellis FW (1886) Tirukkural - English Translation and Commentary. W.H. Allen, & Co, London

Poustie MS, Frantzeskaki N, Brown RR (2016) A transition scenario for leapfrogging to a sustainable urban water future in Port Vila, Vanuatu Technological Forecasting and Social Change 105:129-139 doi:http://dx.doi.org/10.1016/j.techfore.2015.12.008

Pressman RS (2005) Software engineering: a practitioner's approach. Palgrave Macmillan,

Quan NH, Phi HL, Tran PG, Radhakrishnan M, Quang CNX, Thuyen LX, Vinh KQ (2014) Urban retention basin in developing city: from theoretical effectiveness to practical feasibility. Paper presented at the 13th International Conference on Urban Drainage, Kuching, Malaysia,

Radhakrishnan M (2015) Closing the gap between knowledge and practice. Mare Asia. http://mare-asia.net/closing-the-gap-between-knowledge-and-practice/. Accessed 12 Feb 2016

Radhakrishnan M, Ashley R, Gersonius B, Pathirana A, Zevenbergen C (2016) Flexibility in adaptation planning: Guidelines for when, where & how to embed and value flexibility in an urban flood resilience context. CRCWSC - Cooperative Research Centre for Water Sensitive Cities, Melbourne, Australia

Radhakrishnan M, Islam T, Pathirana A, Ashley MR, Nguyen HQ, Gersonious B, Zevenbergen C (Submitted-a) Context specific adaptation grammar (algorithms) for climate adaptation in urban areas Environmental Modelling and Software

Radhakrishnan M, Pathirana A, Ashley MR, Gersonious B, Zevenbergen C (Submitted-b) Flexible adaptation planning for Water Sensitive Cities Cities

Radhakrishnan M, Lowe R, Gersonious B, Ashley R, Arnbjerg-Nielsen K, Pathirana A, Zevenbergen C (Submitted-c) Flexible adaptation planning in a water sensitive Melbourne Proceedings of the Institution of Civil Engineers - Municipal Engineer

Radhakrishnan M, Pathirana A, Ashley R, Zevenbergen C (2017a) Structuring climate adaptation through multiple perspectives: Framework and case study on flood risk management Water 9:129 doi:10.3390/w9020129

Radhakrishnan M, Quan NH, Gersonius B, Pathirana A, Vinh KQ, Ashley MR, Zevenbergen C (2017b) Coping capacities for improving adaptation pathways for flood protection in Can Tho, Vietnam Climatic Change doi:10.1007/s10584-017-1999-8

Ranger N, Millner A, Dietz S, Fankhauser S, Lopez A, Ruta G (2010) Centre for Climate Change Economics and Policy Adaptation in the UK: a decision-making process

Reeder T, Ranger N (2011) How do You Adapt in an Uncertain World? Lessons from the Thames Estuary 2100 Project, World Resources Report.

Revi A, Satterthwaite DE, Aragón-Durand F, Corfee-Morlot J, Kiunsi RBR, Pelling M, Roberts DC, Solecki W (2014) Urban areas. In: Field CB et al. (eds) Climate Change 2014: Impacts, Adaptation, and Vulnerability. Part A: Global and Sectoral Aspects. Contribution of Working Group II to the Fifth Assessment Report of the Intergovernmental Panel of Climate Change. Cambridge University Press, Cambridge, United Kingdom and New York, NY, USA, pp 535-612

Rickard CE (2009) Flood Walls and Flood Embankments. In: Ackers JC, Rickard CE, Gill DS (eds) Fluvial Design Guide. Environment Agency, UK, London,

Rijke J, Ashley MR, Sakic R (2016) Adaptation mainstreaming for achieving flood resilience in cities. CRCWSC - Cooperative Research Centre for Water Sensitive Cities, Melbourne, Australia

Roberts BH (2014) Managing Systems of Secondary Cities - Policy Responses in International Development. City Alliance / UNOPS, Brussels

Rodriguez CS, Radhakrishnan M, Ashley MR, Gersonius B (2016) Extended ATP approach to include the four domains of flood risk managment - Manual with Prototype software tool. CRCWSC - Cooperative Research Centre for Water Sensitive Cities, Melbourne, Australia

Rogers BC, Bertram N, Gunn A, Löwe R, Murphy C, Pasman R, Radhakrishnan M, Urich C, Arnbjerg-Nielsen K Exploring Elwood's flood challenges: A Collaborative Approach for a Complex Problem. In: 2nd Water Sensitive Cities Conference, Melbourne, 2015.

Rogers CD, Lombardi DR, Leach JM, Cooper R (2012) The urban futures methodology applied to urban regeneration Proceedings of ICE Engineering Sustainability 165:5-20

Ross AM, Rhodes DH, Hastings DE (2008) Defining changeability: Reconciling flexibility, adaptability, scalability, modifiability, and robustness for maintaining system lifecycle value Systems Engineering 11:246-262 doi:10.1002/sys.20098

Rossman LA (2010) Storm Water Management Model User's manual - version 5.0. United States Environmental Protection Agency, Cincinaati

Rözer V, Müller M, Bubeck P, Kienzler S, Thieken A, Pech I, Schröter K, Buchholz O, Kreibich H (2016) Coping with Pluvial Floods by Private Households Water 8:304

Salama AM, Wiedmann F (2016) Perceiving urban liveability in an emerging migrant city Proceedings of the Institution of Civil Engineers - Urban Design and Planning 169:268-278 doi:10.1680/jurdp.15.00034

Sallis JF, Bull F, Burdett R, Frank LD, Griffiths P, Giles-Corti B, Stevenson M (In press) Use of science to guide city planning policy and practice: how to achieve healthy and sustainable future cities The Lancet doi:10.1016/s0140-6736(16)30068-x

Samoli E, Atkinson RW, Analitis A, Fuller GW, Green DC, Mudway I, Anderson HR, Kelly FJ (2016) Associations of short-term exposure to traffic-related air pollution with cardiovascular and respiratory hospital admissions in London, UK Occupational and Environmental Medicine doi:10.1136/oemed-2015-103136

Sánchez AM, Pérez MP (2005) Supply chain flexibility and firm performance: A conceptual model and empirical study in the automotive industry International Journal of Operations & Production Management 25:681-700 doi:doi:10.1108/01443570510605090

Satterthwaite D (2007) Adapting to climate change in urban areas: the possibilities and constraints in low-and middle-income nations vol 1. Iied,

Sayers P, Galloway G, Penning-Rowsell E, Yuanyuan L, Fuxin S, Yiwei C, Kang W, Le Quesne T, Wang L, Guan Y (2015) Strategic flood management: ten 'golden rules' to guide a sound approach International Journal of River Basin Management 13:137-151 doi:10.1080/15715124.2014.902378

Sayers P, Yuanyuan L, Galloway G, Penning-Rowsell E, Fuxin S, Kang W, Yiwei C, Le Quesne T (2013) Flood Risk Management : A Strategic Approach. UNESCO, Paris

SCE (2013) Can Tho (Vietnam) : Comprehensive Resilience Planning For Integrated Flood Risk Mangement - Final Report WorldBank,

Schaer C (2015) Condemned to live with one's feet in water?: A case study of community based strategies and urban maladaptation in flood prone Pikine/Dakar, Senegal International Journal of Climate Change Strategies and Management 7:534-551 doi:doi:10.1108/IJCCSM-03-2014-0038

Schulz AP, Fricke E, Igenbergs E (2000) Enabling Changes in Systems throughout the Entire Life-Cycle – Key to Success? INCOSE International Symposium 10:565-573 doi:10.1002/j.2334-5837.2000.tb00426.x

Seakins B (2013) Stawell Steps. http://architectureau.com/articles/stawell-steps/. Accessed 26/10/2016 2016

Serrao-Neumann S, Crick F, Harman B, Schuch G, Choy DL (2015) Maximising synergies between disaster risk reduction and climate change adaptation: Potential enablers for improved planning outcomes Environmental Science & Policy 50:46-61 doi:http://dx.doi.org/10.1016/j.envsci.2015.01.017

Simpson M, James R, Hall JW, Borgomeo E, Ives MC, Almeida S, Kingsborough A, Economou T, Stephenson D, Wagener T (2016) Decision Analysis for Management of Natural Hazards Annual Review of Environment and Resources 41:489-516 doi:doi:10.1146/annurev-environ-110615-090011

Simpson TW, Marion T, de Weck O, Hölttä-Otto K, Kokkolaras M, Shooter SB Platform-based design and development: current trends and needs in industry. In: ASME 2006 international design engineering technical conferences and computers and information in engineering conference, 2006. American Society of Mechanical Engineers, pp 801-810

SIWRP (2011) The Flood Protection Plan for Can Tho city. Southern Institute for Water Resources Planning (SIWRP), Ho Chi Minh City

Smajgl A, Toan TQ, Nhan DK, Ward J, Trung NH, Tri LQ, Tri VPD, Vu PT (2015) Responding to rising sea levels in the Mekong Delta Nature Clim Change 5:167-174 doi:10.1038/nclimate2469

Spaans M, Waterhout B (In Press) Building up resilience in cities worldwide – Rotterdam as participant in the 100 Resilient Cities Programme Cities In Press doi:http://dx.doi.org/10.1016/j.cities.2016.05.011

Spiller M, Vreeburg JHG, Leusbrock I, Zeeman G (2015) Flexible design in water and wastewater engineering – Definitions, literature and decision guide Journal of Environmental Management 149:271-281 doi:http://dx.doi.org/10.1016/j.jenvman.2014.09.031

Stern N (2007) The economics of climate change: the Stern review. Cambridge University press,

Storch H, Downes NK (2011) A scenario-based approach to assess Ho Chi Minh City's urban development strategies against the impact of climate change Cities 28:517-526 doi:http://dx.doi.org/10.1016/j.cities.2011.07.002

Street RB, Nilsson C (2014) Introduction to the Use of Uncertainties to Inform Adaptation Decisions. In: Capela Lourenço T, Rovisco A, Groot A, Nilsson C, Füssel H-M, Van Bree L, Street BR (eds) Adapting to an Uncertain Climate: Lessons From Practice. Springer International Publishing, Cham, pp 1-16. doi:10.1007/978-3-319-04876-5_1

Suh ES, de Weck OL, Chang D (2007) Flexible product platforms: framework and case study Research in Engineering Design 18:67-89 doi:10.1007/s00163-007-0032-z

TE2100 (2012) Managing flood risk through London and the Thames estuary - TE2100 Plan. Environment Agency, London

Tessler ZD, Vörösmarty CJ, Grossberg M, Gladkova I, Aizenman H, Syvitski JPM, Foufoula-Georgiou E (2015) Profiling risk and sustainability in coastal deltas of the world Science 349:638-643 doi:10.1126/science.aab3574

Thiruvalluar (31 BC) Tirukkural. Tamilnadu, India

Thorn J, Thornton TF, Helfgott A (2015) Autonomous adaptation to global environmental change in peri-urban settlements: Evidence of a growing culture of innovation and revitalisation in Mathare Valley Slums, Nairobi Global Environmental Change 31:121-131 doi:http://doi.org/10.1016/j.gloenvcha.2014.12.009

Thorne CR, Evans EP, Penning-Rowsell EC (2007) Future flooding and coastal erosion risks. Thomas Telford,

Toole S, Klocker N, Head L (2015) Re-thinking climate change adaptation and capacities at the household scale Climatic Change 135:203-209 doi:10.1007/s10584-015-1577-x

Triantis AJ (2003) Real options Handbook of modern finance:D1-D32

Tyler S, Moench M (2012) A framework for urban climate resilience Climate and Development 4:311-326 doi:10.1080/17565529.2012.745389

UN (2015) Transforming our world: The 2030 Agenda for Sustainable Development. United Nations, New York

UNEP (2014) The Adaptation Gap Report 2014. United Nations Environment Programme (UNEP), Nairobi

van Buuren A, Driessen PPJ, van Rijswick M, Rietveld P, Salet W, Spit T, Teisman G (2013) Towards Adaptive Spatial Planning for Climate Change: Balancing Between Robustness and Flexibility Journal for European Environmental & Planning Law 10:29-53 doi:doi:http://dx.doi.org/10.1163/18760104-01001003

van der Brugge R, Roosjen R (2015) An institutional and socio-cultural perspective on the adaptation pathways approach Journal of Water and Climate Change 6:743-758 doi:10.2166/wcc.2015.001

van Haegen MS, Wieriks K (2015) The Deltaplan revisited: changing perspectives in the Netherlands' flood risk reduction philosophy Water Policy 17:41-57

Van P, Popescu I, Van Griensven A, Solomatine D, Trung N, Green A (2012) A study of the climate change impacts on fluvial flood propagation in the Vietnamese Mekong Delta Hydrology and Earth System Sciences 16:4637-4649

Veerbeek W, Gersonius B, Ashley R, Radhakrishnan M, Rodriguez CS (2016) Appropiate Flood Adaptation: Adapting in the right way, in the right place and at the right time. CRCWSC - Cooperative Research Centre for Water Sensitive Cities, Melbourne, Australia

Verhoeff M, Verhagen WJC, Curran R (2015) Maximizing Operational Readiness in Military Aviation by Optimizing Flight and Maintenance Planning Transportation Research Procedia 10:941-950 doi:https://doi.org/10.1016/j.trpro.2015.09.048

VIAP-SUIP (2013) Master Plan of Can Tho city until 2030 and Vision to 2050 Southern Sub-Institute of Urban and Rural Planning (VIAP-SIUP), Ha Noi

Victoria (2014) Plan Melbourne- Metropoliton planning strategy. Victorian Government, Melbourne, Victoria

Victoria (2016a) All things considered. Infrastructure Victoria, Melbourne

Victoria (2016b) Draft options book vol Version Two. Infrastructure Victoria, Melbourne

Vink M, Boezeman D, Dewulf A, Termeer C (2014) 3 Action research in governance landscapes Action Research for Climate Change Adaptation: Developing and Applying Knowledge for Governance:35

Von Bertalanffy L (1972) The history and status of general systems theory Academy of Management Journal 15:407-426

Walker WE, Haasnoot M, Kwakkel JH (2013) Adapt or perish: a review of planning approaches for adaptation under deep uncertainty Sustainability 5:955-979

Walker WE, Rahman SA, Cave J (2001) Adaptive policies, policy analysis, and policy-making European Journal of Operational Research 128:282-289 doi:http://dx.doi.org/10.1016/S0377-2217(00)00071-0

Wamsler C, Brink E (2014a) Interfacing citizens' and institutions' practice and responsibilities for climate change adaptation Urban Climate 7:64-91 doi:http://doi.org/10.1016/j.uclim.2013.10.009

Wamsler C, Brink E (2014b) Moving beyond short-term coping and adaptation Environment and Urbanization 26:86-111 doi:doi:10.1177/0956247813516061

Wang T (2005) Real options" in" projects and systems design: identification of options and solutions for path dependency. Massachusetts Institute of Technology

Wang T, De Neufville R (2005) Real options "in" projects 9th Real Options Annual International Conference, Paris, FR

Ward N, Donaldson A, Lowe P (2004) Policy Framing and Learning the Lessons from the UK's Foot and Mouth Disease Crisis Environment and Planning C: Government and Policy 22:291-306 doi:10.1068/c0209s

Wassmann R, Hien N, Hoanh C, Tuong T (2004) Sea Level Rise Affecting the Vietnamese Mekong Delta: Water Elevation in the Flood Season and Implications for Rice Production Climatic Change 66:89-107 doi:10.1023/B:CLIM.0000043144.69736.b7

Wendler R The Structure of Agility from Different Perspectives. In: Computer Science and Information Systems (FedCSIS), 2013 Federated Conference on, 8-11 Sept. 2013 2013. pp 1177-1184

Wesselink A, Warner J, Syed MA, Chan FKS, Tran DD, Huq H, Fredrik H, Ngan Le T, Nicholas P, Martijn VS (2016) Trends in flood risk management in deltas around the world: Are we going 'soft'? International Journal of Water Governance 3:25–46

Wise RM, Fazey I, Stafford Smith M, Park SE, Eakin HC, Archer Van Garderen ERM, Campbell B (2014) Reconceptualising adaptation to climate change as part of pathways of change and response Global Environmental Change 28:325-336 doi:http://dx.doi.org/10.1016/j.gloenvcha.2013.12.002

Wolf J (2011) Climate Change Adaptation as a Social Process. In: Ford DJ, Berrang-Ford L (eds) Climate Change Adaptation in Developed Nations: From Theory to Practice. Springer Netherlands, Dordrecht, pp 21-32. doi:10.1007/978-94-007-0567-8_2

Wong T, Brown R (2009) The water sensitive city: principles for practice Water Science & Technology—WST 60:673–682 doi:10.2166/wst.2009.436

Wong TH (2006) Water sensitive urban design–the journey thus far Australian Journal of Water Resources 10:213-222

Woodward M, Kapelan Z, Gouldby B (2014) Adaptive Flood Risk Management Under Climate Change Uncertainty Using Real Options and Optimization Risk Analysis 34:75–92 doi:10.1111/risa.12088

World Bank (2012) Tools for Building Urban Resilience: Integrating Risk Information into Investment Decisions. Pilot Cities Report – Jakarta and Can Tho. Disaster Risk Management Team, East Asia and Pacific Infrastructure Unit (EASIN), The World Bank,

World Bank (2014) Can Tho, Vietnam Enhancing Urban Resilience : Cities Strength - Resilient Cities program. The World Bank Group, Washington

Young K, Hall JW (2015) Introducing system interdependency into infrastructure appraisal: from projects to portfolios to pathways Infrastructure Complexity 2:1-18 doi:10.1186/s40551-015-0005-8

Zeff HB, Herman JD, Reed PM, Characklis GW (2016) Cooperative drought adaptation: Integrating infrastructure development, conservation, and water transfers into adaptive policy pathways Water Resources Research 52:7327-7346 doi:10.1002/2016wr018771

Zevenbergen C, Rijke J, van Herk S, Bloemen P (2015a) Room for the River: a stepping stone in Adaptive Delta Management International Journal of Water 3:121-140

Zevenbergen C, van de Guchte C, Pathirana A, Wieriks K, N V (2015b) Water and Delta-Cities: Accelerating Urban Resilience Paper presented at the International Conference on Earth Observation & Societal Impacts and ICLEI Resilience Forum, Kaohsiung, Taiwan, ROC, June 28-30

Zhang SX, Babovic V (2012) A real options approach to the design and architecture of water supply systems using innovative water technologies under uncertainty Journal of Hydroinformatics 14:13–29 doi:10.2166/hydro.2011.078

Zhang Z, Sharifi H (2000) A methodology for achieving agility in manufacturing organisations International Journal of Operations & Production Management 20:496-513 doi:doi:10.1108/01443570010314818

About the Author

Mohanasundar Radhakrishnan hails from an agrarian society in Tamil Nadu, India. He obtained his Bachelor's degree in Civil Engineering from University of Madras in 2002 and MSc. degree in Municipal Water and Infrastructure from IHE Delft in 2009. He worked as a design engineer and was involved in the hydraulic design of drinking water distribution networks and bulk water transmission main in various water supply schemes in India from June 2002 to Oct 2007. After his masters from IHE Delft he worked with Arghyam, an NGO as a Project officer in an integrated urban water management project in Karnataka, India till 2010. He worked as Project Advisor for German International cooperation (GIZ) and contributed towards the preparation of City sanitation Plans and planning of a waste to energy project in Nashik, India. He is a registered volunteer with Mediciens Sans Frontier (MSF) and served as a Water and sanitation specialist in Loas towards improving the water, sanitation and waste disposal facilities of five hospitals in rural areas between Dec 2012 and May 2013. He is now associated with IHE Delft's Flood resilience chair group of Water Science and Engineering Department as a full time PhD student, researching on embedding flexibility in Urban Flood Risk Management systems. His research was funded by Govt. Of Australia through CRC for Water sensitive cities, a research initiative which brings together the inter-disciplinary research expertise and thought-leadership to undertake research that will revolutionise water management in Australia and overseas.

Publications by the author

Journal Papers

Radhakrishnan M, Pathirana A, Ashley R, Zevenbergen C (2017) Structuring climate adaptation through multiple perspectives: Framework and case study on flood risk management Water 9:129 doi:10.3390/w9020129

Radhakrishnan M, Quan NH, Gersonius B, Pathirana A, Vinh KQ, Ashley MR, Zevenbergen C (2017) Coping capacities for improving adaptation pathways for flood protection in Can Tho, Vietnam Climatic Change doi:10.1007/s10584-017-1999-8

Radhakrishnan M, Ashley R, Gersonius B, Pathirana A, Zevenbergen C (Submitted) Flexible adaptation planning process for water sensitive cities Cities

Radhakrishnan M, Islam T, Pathirana A, Ashley R, Nguyen HQ, Gersonious B, Zevenbergen C (Submitted) Context specific adaptation grammar (algorithms) for climate adaptation in urban areas Environmental Science and Modelling

Radhakrishnan M, Lowe R, Gersonious B, Ashley R, Arnbjerg-Nielsen K, Pathirana A, Zevenbergen C (Submitted) Flexible adaptation planning in a water sensitive Melbourne Proceedings of the Institution of Civil Engineers - Engineering Sustainability

Radhakrishnan M, Pathak TM, Irvine K, Pathirana A (Submitted) Scoping for the operation of agile urban adaptation for secondary cities of global south: Possibilities in Pune, India Water 9: xxx

Radhakrishnan M, Zevenbergen C, Pathirana A, Ashley R (Submitted) The fourth generation climate adaptation challenge: Implementing adaptation responses Environmental Science & Policy

Goh XP, Radhakrishnan M, Zevengergen C, Pathirana A (2017) Effectiveness of Runoff control legistation and and Active, Beautiful and Clean (ABC) Waters Design Features in Singapore Water 9: 627

Nguyen HQ, Radhakrishnan M, Huynh TTN, Baino-Salingay ML, Ho LP, Steen PVd, Pathirana A (2017) Water Quality Dynamics of Urban Water Bodies during Flooding in Can Tho City, Vietnam Water 9:260

Pathirana A, Radhakrishnan M, Quan NH, Zevenbergen C (2017) Managing urban water systems with significant adaptation deficits – unified framework for secondary cities : part I - conceptual framework Climatic Change

Pathirana A, Radhakrishnan M, Ashley R, Quan NH, Zevenbergen C (2017) Managing urban water systems with significant adaptation deficits – unified framework for secondary cities : Part II – the practice Climatic Change

Vincet SU, Radhakrishnan M, Hayde L, Pathirana A, (Accepted) Enhancing the economic value of large investments in Sustainable Drainage Systems (SuDS) through inclusion of Ecosystem Service benefits Water 9: xxx

Yau WK, Radhakrishnan M, Liong SY, Pathirana A, Zevengergen C (2017) Effectiveness of ABC Waters Design Features for Runoff Quantity Water 9: 577

Radhakrishnan M, Pathirana A, Ghebremichael K, Amy G (2012) Modelling formation of disinfection by-products in water distribution: optimisation using a multi-objective evolutionary algorithm Journal of Water Supply: Research and Technology—AQUA 61:176-188

Pathirana A, Gersonius B, Radhakrishnan M (2012) Web 2.0 collaboration tool to support student research in hydrology–an opinion Hydrology and Earth System Sciences 16:2499-2509

Conference papers

Radhakrishnan M, Pathirana A, Gersonius B, Zevenbergen C, Ashley R (2013) Resilience approach to Urban Flood Risk Management systems using Real in Options - a review. Paper presented at the Water Sensitive Urban Design Gold Coast,

Radhakrishnan M, Quang CNX, Pathirana A, Phi HL, Quan NH, Ashley MR (2014a) Evaluation of Retrofitting Options in Urban Drainage Systems based on Flexibility: A Case Study For Nhieu Loc - Thi Nghe Basin in Ho Chi Minh City. Paper presented at the HIC 2014 – 11th International Conference on Hydroinformatics "Informatics and the Environment: Data and Model Integration in a Heterogeneous Hydro World", New York, USA,

Radhakrishnan M, Quang CNX, Pathirana A, Phi HL, Quan NH, Ashley RM (2014b) Retrofitting urban drainage capacity to cope with change: A case study for Nhieu Loc - Thi Nghe Basin change in Ho Chi Minh city. Paper presented a-t the 13th International Conferece on Urban Drainage 2014, Sarawak, Malaysia,

Lowe R, Urich C, Radhakrishnan M, Deletic A, Arnbjerg-Nielsen K Scenario-free Simulation of Flood Risk for Multiple Drivers. In: Seventh International Conference on Flood Managemet (ICFM), Leeds, UK, 2017. University of Leeds,

Quan NH, Phi HL, Tran PG, Radhakrishnan M, Quang CNX, Thuyen LX, Vinh KQ (2014) Urban retention basin in developing city: from theoretical effectiveness to practical feasibility. Paper presented at the 13th International Conference on Urban Drainage, Kuching, Malaysia,

Rogers BC, Bertram N, Gunn A, Löwe R, Murphy C, Pasman R, Radhakrishnan M, Urich C, Arnbjerg-Nielsen K Exploring Elwood's flood challenges: A Collaborative Approach for a Complex Problem. In: 2nd Water Sensitive Cities Conference, Melbourne, 2015.

Reports

Radhakrishnan M, Ashley R, Gersonius B, Pathirana A, Zevenbergen C (2016) Flexibility in adaptation planning: Guidelines for when, where & how to embed and value flexibility in an urban flood resilience context. CRCWSC - Cooperative Research Centre for Water Sensitive Cities, Melbourne, Australia

Rodriguez CS, Radhakrishnan M, Ashley MR, Gersonius B (2016) Extended ATP approach to include the four domains of flood risk managment - Manual with Prototype software tool. CRCWSC - Cooperative Research Centre for Water Sensitive Cities, Melbourne, Australia

Veerbeek W, Gersonius B, Ashley R, Radhakrishnan M, Rodriguez CS (2016) Appropiate Flood Adaptation: Adapting in the right way, in the right place and at the right time. CRCWSC - Cooperative Research Centre for Water Sensitive Cities, Melbourne, Australia

Gersonius B, Ashley R, Salinas Rodriguez CNA, Rijke J, Radhakrishnan M, Zevenbergen C (2016) Flood Resilience in Water Sensitive Cities. Cooperative Research Centre for Water Sensitive Cities, Clayton, Melbourne, Australia

Netherlands Research School for the
Socio-Economic and Natural Sciences of the Environment

DIPLOMA

For specialised PhD training

The Netherlands Research School for the
Socio-Economic and Natural Sciences of the Environment
(SENSE) declares that

Mohanasundar Radhakrishnan

born on 19 March 1981 in Palayamkottai, India

has successfully fulfilled all requirements of the
Educational Programme of SENSE.

Delft, 22 November 2017

the Chairman of the SENSE board

Prof. dr. Huub Rijnaarts

the SENSE Director of Education

Dr. Ad van Dommelen

The SENSE Research School has been accredited by the Royal Netherlands Academy of Arts and Sciences (KNAW)

KONINKLIJKE NEDERLANDSE
AKADEMIE VAN WETENSCHAPPEN

The SENSE Research School declares that **Mr Mohanasundar Radhakrishnan** has successfully fulfilled all requirements of the Educational PhD Programme of SENSE with a work load of 38.6 EC, including the following activities:

<u>SENSE PhD Courses</u>

o   Environmental research in context (2013)
o   Research in context activity: 'Co-organising of international workshop for practitioners and stakeholders on "Developing climate adaption and green infrastructure in medium-sized cities across multiple scales", Vietnam' (2014)
o   SENSE writing course (2016)

<u>Other PhD and Advanced MSc Courses</u>

o   Water resilient cities, UNESCO-IHE (2014)
o   Communicating water - bridging the gap between science and society, UNESCO-IHE (2015)

<u>Management and Didactic Skills Training</u>

o   Member and Chair of the PhD Association Board (PAB) of UNESCO-IHE (2014-2015)
o   Supervising two MSc students with thesis entitled 'Broader environmental impact of runoff control using SuDS in the context of urban development in Singapore' (2015) and 'Flexibility assessment of urban drainage solutions in developing cities - a case study in Can Tho, Vietnam' (2015)
o   Member of multi-disciplinary Elwood climate adaptation research group (2016)

<u>Oral Presentations</u>

o   *Resilience approach to urban flood risk management systems using Real in Options.* Water Sensitive Urban Design (WSUD), 25-29 November 2013, Gold Coast, Australia
o   *Effect of coping capacities on enhancing adaptation tipping points.* Deutscher Kongress für Geopgraphie (DKG), 3 October 2015, Berlin, Germany
o   *Improving flood resilience of Can Tho city: Big Infrastructure.* Green infrastructure for water sensitive urban development in secondary cities of Global South, 8 December 2015, Can Tho, Vietnam

SENSE Coordinator PhD Education

Dr. ing. Monique Gulickx

Printed and bound by CPI Group (UK) Ltd, Croydon, CR0 4YY

22/10/2024

01777333-0005